SKY
STATIC

More than 8,000 spacecraft have been put into orbit since 1957, and up to 140 per year are now launched. Some perform delicate retrieval or docking maneuvers, but most become pitted or even disabled by space debris. Electronic systems are particularly vulnerable to cosmic rays. Photo courtesy of NASA.

SKY STATIC

The Space Debris Crisis

Antony Milne

Westport, Connecticut
London

Library of Congress Cataloging-in-Publication Data

Milne, Antony.
   Sky static : the space debris crisis / Antony Milne.
      p. cm.
   Includes bibliographical references and index.
   ISBN 0–275–97749–8 (alk. paper)
   1. Space debris. I. Title.
   TL1499.M55   2002
   363.728'0919—dc21   2002067293

British Library Cataloguing in Publication Data is available.

Library of Congress Catalog Card Number: 2002067293
ISBN: 0–275–97749–8

First published in 2002

Praeger Publishers, 88 Post Road West, Westport, CT 06881
An imprint of Greenwood Publishing Group, Inc.
www.praeger.com

Printed in the United States of America

The paper used in this book complies with the
Permanent Paper Standard issued by the National
Information Standards Organization (Z39.48–1984).

10 9 8 7 6 5 4 3 2 1

Contents

Contents

Acknowledgments

My thanks go to many people who have given me experience of their knowledge—especially in the field of meteorites, satellite construction and tracking, the workings of the European Space Agency (ESA), and space scientists generally, many of whom have read parts of the manuscript. They include Phillip Clark of the Molniya Space Consultancy, Donald Kessler and Walter Flury of the ESA, Jonathan Tate of Spaceguard UK, Dr. D.J. Gardner of the Space Science Dept of Kent University, Richard Crowther of DERA (Qinetiq) in Farnborough, England, and Dr. A.D. Romig of the Sandia National Laboratories.

Introduction

This book appears as a tribute to the growing consultancy and committee work, especially at the European level, of a wide range of space
and defense scientists, amounting now to many hundreds of experts,
who are quietly tackling what they consider to be a new environmental menace: space debris. Those who attended the enormous
gatherings of scientists in recent years in Darmstadt, Germany ("City
of Science and Research") realize how interdisciplinary and complex
the subject of space debris has become in the post–Cold War world—
involving applied space science in a mammoth way—including measurement, radar and telescope tracking, debris experiments, meteorite
dust analysis, materials science, modeling, shield design, risk analysis,
mitigation policies, as well as defense policy, legal, and international
issues.

Indeed it is the risk aspect that has spurred me to write this book.
One could argue that too many international gatherings concerning
the fate of the world have been high on rhetoric and exhortation and
a bit low in regard to the actual science base. With the space debris
threat it is the opposite: debris crashing to Earth has actually made
the headlines.

Those of us who have given talks on the subject of the threat of meteorite impacts have often been disappointed at the rather skeptical way in which the subject is received. The threat of major harm to populations as a result of a half-mile-wide meteor crashing to Earth is dismissed as being too hypothetical or having odds so extremely long that it is not worth bothering about.

My own position was shifted away to space debris as a more probable and imminent threat in early 1996, when the world's press, for almost a week, reported the antics of the out-of-control Chinese military satellite, weighing several tons, which could have crash-landed anywhere on Earth with incalculable consequences. There was also the publicity given to the giant tank-like objects and rocket component parts that fell one-by-one onto South Africa in 2000, and the burn-up of the defunct *Mir* space station in the spring of 2001, which illuminated Asian skies. Although *Mir* was successfully guided to its seabed graveyard in the South Pacific, it was clear that natural radio gaps in communication, the impact of space radiation, or simple malfunction, means that deorbiting satellites successfully to avoid striking populated areas is by no means guaranteed, as this book will make clear.

The main concern about space debris is admittedly economic: the damage that is likely to occur to the existing and growing number of commercial and science satellites in space or in the process of being sent there. Satellites are extremely expensive and need to be protected from their own cluttering effects. There is the "cascading" effect, too. When a satellite is detroyed either by a collision with natural or man-made debris, it can add to the flotsam already in space. Broadcasting signals can be blotted out, TV pictures can be interrupted, and vital intelligence can be lost to spy satellites simply by a zapping stray cosmic ray. There is a lot at stake in the space business, which is now almost entirely a commercial venture. The multimedia companies naturally insist on space safety, and so do the military. And we on the surface, it seems, also need some guarantee that our homes and cities will not be at risk from space junk. As recent international events have shown, we can no longer be complacent about anything anymore. Hollywood disaster movies have a way of coming true with a vengeance.

CHAPTER 1

The *Mir* Fireball

Late in the morning of 23 March 2001, in the eastern hemisphere, a man-made crisis occurred that had all the hallmarks of a natural disaster combined with a terrorist attack. A massive 140-ton object—the size of a medium-sized meteorite—was about to crash onto Earth. It was only a matter of luck that the object, fragmenting into thousands of pieces to spread over an area of hundreds of square miles, did not land on populated areas to cause widespread death and destruction.

But there was another important aspect to this extraordinary event. This was the first time aeronautic scientists had no alternative but to purposely create *space debris*—and space debris is now regarded as a new and growing environmental hazard. It was seen as inevitable that the Russian space station *Mir* would sooner or later become space junk. All satellites and space stations have to be deorbited at some time, and will come down eventually when their fuel runs out, unless of course they are partaking in an interplanetary mission and are way beyond the pull of Earth's gravity. Then they will simply be lost in space forever. Thus space junk from dead Earth-orbiting satellites is a virtual certainty.

Indeed five satellites a week reenter the atmosphere and mostly burn up, and often large chunks of spent rocket boosters and the more solid parts of spacecraft actually reach the ground (see Chapter 6).

Mir had at last, and some say not before time, come to the end of its long working life. It was originally planned to stay in space for about four years, but in fact lasted 15. Indeed Professor Andre Balogh, a space physicist from Imperial College, London, feared *Mir* might veer out of human control, so decrepit had its communication system, its infrastructure housing and its added-on modules—not to mention its engines—become.

There were some alarming moments that suggested this was happening. Although radio link-ups were not intended to be continuous—used only to pass on rocket-firing and steering commands—the Russians had lost touch with the space station for 24 hours on Christmas Day 2000. Mission control managed to restore links for seven minutes on the following day after Christmas and struggled to continue to maintain continuous contact until the very last hour, when *Mir* finally plunged to Earth. Fortunately there was no loss of pressure inside the craft, and temperature and power systems seemed normal, calming fears that the craft, unbeknown to the space scientists, was indeed spinning out of control.

Yet it was worrying that the engineers had no idea what caused the breakdown, although loss of radio contact normally signals a problem with an onboard computer. The Russian space scientists believed they had been in the space-station business long enough to know how to bring down an obsolete satellite. Many of them, especially those at mission control at Korolyev, outside Moscow, scoffed at the alarmist pronouncements being voiced at the beginning of March 2001 that *Mir* was about to blindly crash-land somewhere. Sergei Gorbunov, a spokesman for the space program, said, "It is 100 percent impossible for the station's debris to rain on inhabited areas."[1]

Colonel Norman Black of the U.S. Space Command agreed: it would be easier to win the California lottery than to be hit by a piece of space junk, he declared. William Ailor, director of space debris studies for the Aerospace Corporation in El Segundo, California, dismissed the threat to humans of infalling debris. The risk would be one in a trillion, he said.[2]

In the meantime, from the New Year of 2001 onward, Major Scott Edwards and his number-crunching crew of military rocket scientists, huddled inside a granite bunker 2,000 feet deep in the Rocky Mountains, were taking stock of the 140-ton lump of imminent space debris as it moved across the skies at an ever-quickening pace. They were participating in a joint Russian-American *Mir*-descent monitoring venture. NORAD, the North American Space Defense Command, was now tracking the global network of satellites and radars that warn of ballistic missile attacks,[3] but since the end of the Cold War this had been expanded to cover civilian satellites.

The irony was not lost on Major Edwards in the bomb-proof, quake-proof complex that houses the American military's high-tech missile warning and space command centers, which had been designed as a Cold War era defense citadel and was now lending the United States global surveillance facilities to a former mortal enemy.

Even within America there was sharing of data with the various space and military establishments. Data from the National Aeronautics and Space Administration (NASA) was correlated with the Pentagon's Military Command Center, where it is often shared with the White House and other interested agencies, including the Federal Emergency Management Agency, the State Department, and the Federal Aviation Administration (FAA).

But *Mir*'s reentry required something more than sharing telescopes among astronomers or practicing docking in near space with apolitical civilian spacemen. Anything relevant to *Mir*'s reentry, such as magnetizing solar flares, or other space junk both natural and man-made, was being forwarded to NASA, the main civilian space agency that was liaising with the Russians, to be passed on to *Mir*'s controllers for further analysis and to aid future space missions, some of them bound to have military implications.

Edwards' team was also needed to fill in the gaps in Russian knowledge, since the Russians had monitoring sensors only on Russian territory, whereas the Americans had their sensors and tracking devices spread across the world. Edwards' crew were also monitoring 8,300 other man-made objects—and had additional monitoring input supplied by Britain's Fylingdales (missile) early warning station in Yorkshire.

The Russians had decided it was no longer worthwhile keeping *Mir* up in orbit and fully functioning. They were running out of cash, and they were being pressured by the Americans into redirecting their space science prowess into helping to build the new International Space Station. But bringing down *Mir* posed a dilemma for the Americans. They were keen to help, but wary of the fallout should anything go horribly wrong and people on the surface be pelted with dented hardware. "This isn't an exact science: it's very much an art form," said Colonel Norman Black of the U.S. Space Command.[4]

There were indeed many dissenting voices on both sides. Even Sergei Gorbunov expressed confidence that was not entirely shared by all Russian space engineers. At the end of June 1998, 20 engineers from the Energiya Corporation, the main contractor for *Mir*, complained of lack of funds needed to make the operation a success. It was possible *Mir* could tumble in unexpected ways, or its engines could misfire. "A lot of things can happen in the lower atmosphere that you can't predict," admitted Norman Black.

Major Edwards likened the reentry exercise to skimming a pebble across a pond. The craft could bounce along inside the varying layers of the atmosphere so that precise prediction of its final resting place would be largely calculated guesswork. Nor was that all. Vladimir Lobachyov, the ground control chief, told the Interfax news agency that *Mir* would disappear from radar screens around the world during the last 40 minutes of its life. And this would have nothing to do with a computer malfunction. Even NASA would not be able to monitor its final descent.[5] There was no way of tracking the debris once it entered the Earth's atmosphere, leaving this 40-minute window of uncertainty, largely because of loss of radio contact and even loss of other electronic and visual contact from other satellite monitoring sources. In order to bring *Mir* down safely, the Russian Aviation and Space Agency even took out insurance for £200 million in case it catastrophically crashed into populated areas.

Some years earlier space experts were hinting that they were not at all confident about deorbiting *Mir* successfully. Dr. Richard Crowther, a space expert at the Defense Evaluation and Research Agency (DERA) at Farnborough, England, said it was not going to be an easy operation at all since *Mir* was not designed to reenter the atmosphere because of its odd insect-like shape.[6]

He also pointed out that, unlike other satellites, *Mir* was a manned station and thus more solidly built—a virtual metallic hotel. "The first thing we expect is that *Mir* will line up with the heaviest part at the bottom, like a conker on a string. This is the result of the effect called gravity gradient. Then it will start spinning on its axis, and tumbling."[7]

Further, it would be difficult to know which modules and components would detach themselves first, and—if they did—which ones would resist friction and fail to burn up in Earth's lower atmosphere and thus remain as chunks of solid metal that would reach Earth with a deadly velocity of thousands of miles an hour.

Dr. Crowther said that, based on past experience, *Mir* would break up into at least five major components, all in the 30-ton range, which would burn up separately. Andre Balogh added, "When it comes down it will be in the form of several double-decker bus-sized chunks hurtling down at several hundreds of miles an hour. Clearly it could do an awful lot of damage. . . . We cannot exclude the possibility that it could fall on a major city."[8]

Dr. Crowther also pointed out that around the Christmas period of 1999, just one year before the final decision was taken to deorbit, a solar maximum event would occur, resulting in the dissemination to Earth of high levels of solar radiation that would heat the atmosphere and make it expand. This pushed the upper levels of the atmosphere further upwards and made *Mir* reenter the atmosphere sooner than expected. And it was the atmosphere, with its dragging, friction-inducing properties, that would contribute to the uncertainties about the effects of reentry.

On the other hand, if radio contact was lost, *Mir*'s solar panels could not be pointed accurately at the sun. The station would lose power as the batteries drained. The spacecraft itself would normally drop more than a mile a day if no corrections were made by mission control. Loss of control could lead to a spin, which would make the craft impossible for cosmonauts to start a rescue operation to dock with *Mir* to correct the problem.

The Australians were naturally worried about *Mir*'s debated reentry flight path. "It's fair to say it's taken a few people by surprise," said a spokesman for the Australian Defense Department. Opinion was divided. Some officials estimated there was a one in 5,000 chance that a lump the size of a car might hit their country. Peter Dyson, a space scientist at La Trobe University in Melbourne, said that given the length of the target area, "it seems very likely that particles of *Mir* will enter over Australia."[9]

Walter Flury, a space debris expert at the European Space Agency headquarters in Darmstadt, Germany, pointed out that the trajectory wouldn't actually traverse Australian skies. But he acknowledged the uncertainty about *Mir*'s exact trajectory. The sheer volume of *Mir* debris that would likely survive the heat of reentry was indeed worrying. It was just as well that the South Pacific—its projected "target" area—was extremely wide, geographically speaking, because officials reckoned that about 1,500 fragments weighing as much as 1,600 pounds each would be scattered across a zone of 1,800 miles, a distance half as long again as original predictions.[10] Other estimates suggested that up to 25 tons of nuts, bolts, cannisters, flanges, and assorted metal parts—many still glowing in a semimolten condition—would inevitably come raining down somewhere to make actual contact with Earth's surface.

There was a fear that the six-module station could fall to Earth more steeply than expected, bringing its heaviest components down in the western Pacific. Or it could skid through the atmosphere and head further toward southern Chile. The nearest named land for the assumed impact zone was the uninhabited Maria Theresa Reef, 1,500 miles south of Tahiti. Rosiaviacosmos, the Russian space agency, tried to reassure the Tongan and Fijian governments. You cannot, it seems, precisely target infalling space junk, especially as the atmospheric burn-up will inevitably leave a trail of glowing debris stretching over hundreds of miles. The distance between the forward pieces of debris and the last pieces can never be accurately calibrated. During reentry the front end of the spacecraft would set up heat-creating shock waves.

Of course, since *Mir* would come down somewhere in the Asia-Pacific area, the Japanese authorities tried to reassure nervous citizens. In fact, public concern had been greatest in Japan, despite Moscow's insistence that only the southern tip of the country lay below *Mir*'s reentry path. Toshitsugu Saito, head of Japan's defense agency, postponed a trip to Washington to take charge of "crisis management," as he put it.[11] In New Zealand the authorities restricted themselves to warning shipping and aircraft in the area, which is used by deep-sea fishing vessels and planes flying between Auckland and Los Angeles. In the meantime those on the projected route of the disintegrating *Mir* were warned to stay indoors, just in case, and airline flights were rerouted.

Whatever doubts he had about bringing down *Mir*, Vladimir Solovyov of Rossiaviacosmos believed there could now be no delay in doing it. He admitted that none of *Mir*'s mounting costs and maintenance difficulties could be put right. Moscow itself was hoping for a trouble-free splashdown in the southern Pacific somewhere between Australia and Chile.

SCHEDULING THE FIREBALL IMPACT

The general consensus about *Mir*'s final trajectory and ultimate earthly resting place went something like this: In April 1997 *Mir* was nudged upward for the last time by the Progress M-34 supply craft, a module especially designed for the purpose. It gently descended from 240 miles by August 1998. It was still at this height in July 1999. The last of the main experimental science crew (those who were not simply there as virtual satellite pilots) left the craft.

Since December 2000 the flight controllers had lowered *Mir*'s orbit below 200 miles. When down to 150 miles—still technically at a safe orbiting height—they fired thruster rockets. In January 2001 one of *Mir*'s Progress supply vessels was detached from the space station and plunged into the Pacific as a rehearsal for *Mir*'s descent.

Then a crewless Progress MI-5, with extra fuel, docked with the still-inhabited *Mir* on 22 January to provide the necessary thrust for actual reentry. It started to push *Mir* towards Earth and toward the start of the first layer of the atmosphere, known as the Mesosphere. Then a final series of "burns" took place that were no longer concerned with atmospheric descent, but were used to "target" the craft to a specific landing spot in the South Pacific.

The first burn took place somewhere over Sierra Leone and lasted some several minutes, taking the craft over Iraq, when the burn ended. Then the last cosmonaut aboard the *Mir*, Sergei Avdeyev, steered the craft down to an altitude of 125 miles before climbing into his Soyuz-TM escape capsule in early March 2001 and returning to Earth. Avdeyev had nudged the craft into an elliptical orbit, making it careen—in its death throes—over 80 percent of the Earth's surface, from as far north as London and as far south as the Falkland Islands.

On 12 March, mission control started firing Progress's engines to trigger reentry into the mesosphere proper. The space station did several hundred more orbits of the Earth, slowly sinking over the remaining days of its life. Then the craft was put into its final irreversible decline in a final 23-minute burn by all nine Progress engines. Once this was done there was no turning back. From then on *Mir* had to be controlled all the way down until radio contact could be maintained no longer.

On its very last lap the trajectory continued to swing the craft northwards into a parabolic sweep that took *Mir* across the Caucasian region before lunging down over China, Japan, and into the Melanesian and Polynesian regions of the Pacific Ocean, high above the myriad inhabited islands in the region.

At 0507 hours Greenwich Mean Time (GMT) on 23 March 2001 *Mir* was over Egypt, and at 0535 it passed above Japan. At 0530 it descended from a 110-mile-high final orbit, sinking fast over East Asia. At 60 miles high, aerodynamic forces first ripped off its outer antennae and solar panels. Mission control was now no longer able to issue any further commands. The remaining structure became red hot, soon to be ripped off and vaporized in the 3,000°F degree heat generated through atmospheric friction. At 0544 *Mir* started, finally, to break up. As large parts of the six main pressurized modules began to disassemble, they put on a spectacular display for the lucky few. For those positioned in strategic positions along the way—central Europe, the Caucasion region, the Pacific Islands—the station gave a brief daytime colorful show when streaks of light could be seen in the night skies. Observers heard a series of muffled sonic booms as the space station's remains broke the sound barrier in reverse, slowing in minutes from five miles a second to a few hundred miles an hour.[12]

Mir fragments blazed toward the target area. At about 50 miles up the remaining exterior equipment and body actually came apart. Some 40 propellant tanks, many large batteries, metal storage boxes, and heavy metal bulkheads survived reentry. Then, continuing on its journey, at about 37 miles up, what remained of the structural elements started to spread out into a large area of space. When as low as between 24 and 6 miles high, debris came blistering down over a wide area of between 3,400 and 370 miles.

At 0550 fragments were seen falling near Fiji. Tourists and space buffs looking up from their beaches gasped to see the space station's modules reduced to blazing meteors, torn apart but hurtling in formation. TV footage showed palm leaves bending in the foreground. Some adventurers had paid £4,500 each for seats on three aircraft charted to fly near the target area. Marc Herring, director of Herring Media Group, said minutes before *Mir* broke up he would call Russian mission control to check reentry coordinates, and dodge out of the way if necessary. Among the space customers was Sergei Avdeyev, who spent 748 days on board *Mir*, breaking all records for manned flight. In Tonga local people were hoping to make a quick buck by claiming on the insurance, or selling off any debris they recovered as souvenirs. The Russians were appalled at such insensitivity toward what they saw as one of the saddest epochs in their space program, once the most prestigious in the world.

Mir erupted into a fireball over the South Pacific at 0558 GMT, and the bulk of the debris—some 27 tons of it—landed at that time, just two minutes before 9 AM Moscow time, on Friday 23 March. Although safely within the South Pacific, it ended up some thousand or so miles southeast of New Zealand, somewhat further south than planned, virtually dead center between New Zealand and the coastline of Chile. This fairly wide margin of error validated the cautious concern of space scientists.

A few people in Mission Control cried, while others posed for photos and shook hands. As the success of the splashdown became obvious, Russian officials, such as Yuri Koptev, director of the Russian space agency, exulted in their achievement. "Russia will remain a great space power," he said.

CHAPTER 2

Hazardous Space Missions

We owe the Russian space engineers a debt of gratitude. *Mir* had lasted much longer than they intended, and when the craft finally became space debris, its remnants were skillfully aimed at an uninhabited space on Earth. A horrendous incident was mercifully avoided.

However, many alarming events happened while *Mir* was orbiting for its 15 years that have made space debris experts and aeronautics experts worry. In all the space laboratories sent up over the last 40 years, accidents have abounded. Every one of them has created space debris while still in orbit, as broken component parts from the skin of the various modules and pieces of solar arrays and antennae have been ripped off through accidents or simply through relentless aerodynamic forces.

The orbiting space stations are, technically, "satellites." The difference between space stations and commercial satellites ("comsats") and science-based "probes" is that the latter are naturally smaller and are despatched as one unit, neatly packed into the top stage of a rocket, whereas the former are launched incrementally in modular parts.

The lifting rockets themselves also have the potential to create space debris since many of them fail to properly launch their payloads into correct orbits. And of course the booster stages of rockets fall away into the atmosphere. Many rockets or launchers simply blow up after the first few minutes of flight to bring down tons of metal to the surface (but as they are launched largely from coastal sites, the debris usually, but not always, lands in the ocean).

To a large extent, space debris is the inevitable result of astonishing postwar scientific, commercial, and military achievements. Industrial technology in general has always had its downside, either in terms of the horrendous fatal accidents that occur from time to time, or because of other, mainly sociological, disbenefits that have arisen. Accidents and disasters as by-products of transport technology in particular—involving cars, trains, airplanes, and spacecraft—seem to come at a fearful price, but a price that many feel nevertheless has to be paid.

In terms of space ventures, for both good and ill, the year 1957 was a landmark, since three vital and interrelated developments took place. The first Russian ballistic missile was test-launched in August, shortly before *Sputnik* was despatched into space aboard a modified German wartime V2.

Secondly, with the advent of *Sputnik* itself, the Soviet Union announced to an astonished world that it had launched a rocket with a range of 4,000 miles, on top of which was a capsule containing a "cosmonaut" who would circle the Earth in near space. The focus of analysis, as one defense writer pointed out, was the rocket, not the satellite. The Soviets had made a supreme leap because not only did they have the nuclear bomb, they had a rocket that could carry that bomb into ballistic orbit.[1] Thirdly, *Atlas*, launched in the same year as *Sputnik* and designed to match the new Soviet space weapon, became the first U.S. ballistic missile.

The Americans, by now, had already announced they would develop and launch a satellite in the International Geophysical Year of 1957–58 (a "year" which, strangely, has never been repeated). This gave Wernher von Braun, who was head of a rival project, the opportunity to successfully launch his *Explorer-1* on its Vanguard rocket in January 1958.[2] Events gathered pace quickly. In May the Soviets launched *Sputnik-3*, declaring it to be a scientific laboratory. Yet this, even more so than *Sputnik-1*, was perceived to have military potential. Developments in postwar aerospace technology reflected the respective military capabilities of the emerging superpowers, which would have increasing repercussions on events in international airspace. A need for the research activity to be more focused was discerned, and President Dwight Eisenhower created a special agency, NASA.

Earlier, the great American rocket pioneer, Robert Goddard, set up an experimental station near Roswell, New Mexico. There he developed combustion chambers and perfected liquid fuel burning, cooling, and guidance techniques, including the use of gyroscopes. At the time NASA was formed, improved rocket motors—which the army had developed, using liquid hypergolic, self-igniting fuels—were being used.

But so urgent was the need to press ahead with a space program that when repeated failures occurred with the Vanguard series in the late 1950s and 1960s, converted missiles were pressed into service. The first Anti-ballistic Missiles (ABMs), albeit crude and unreliable, like the Rascal, were designed and built to help older bombers attack ground targets without having to penetrate enemy airspace. Later the Thor rocket series was an Intermediate Range Ballistic Missile (IRBM) developed by the air force at the same time as the Jupiter IRBM was being built by the army.[3]

Military and aerospace institutions began to proliferate and flourish, and wartime establishments were refurbished and reequipped. The connection between the military and between space research and rocket design was becoming more obvious, and various aeronautics and launch components were swapped and cross-used prior to the Atlas and Vanguard programs. White Sands in New Mexico was the site from which the United States originally fired most of the V2 rockets captured from the Germans, and its administrators soon received new contracts for new rockets.

Other establishments, like the Marshall Space Flight Center, started a new recruitment drive for technicians and space scientists. The center, named after General of the Army George C. Marshall, is in the midst of the U.S. Army Redstone Arsenal at Huntsville, Alabama, which was the successor to the Army Ballistic Missile Agency, famous for developing the Jupiter series of rockets.[4]

Most rocket theorists thought that the entire mechanical edifice would have to reach its destination intact. The early equations seemed to demand that up to 90 percent of launch weight must be the fuel itself. Goddard, however, formulated the principle of the multistage rocket. The first stage burned for a predetermined time to lift the launch vehicle through the lowest and densest part of the atmosphere.

After it stopped firing, the second stage took over, and explosive charges inside the bolts holding the spent stage in place fired, and it then fell away under the pull of gravity.

Each successive stage shed part of the original weight, so the remainder of the space vehicle started at the higher velocity imparted by the preceding stage, and thus needed less fuel to reach that velocity. Occasionally small retrograde rocket motors mounted around the top fired to slow the jettisoned stage down to prevent it from striking the still-climbing launch vehicle.[5] Gyroscopes kept the vehicle spinning around its original axis and "remembered" its direction of spin, thus keeping the rocket on a level course. Onboard mechanical and electronic devices detected and measured any departure from the planned course. Later rockets had computers that worked out what corrections had to be made and then varied the angle of thrust produced by the motors, which were designed to pivot and tilt during flight. This in turn meant that the propellant pipeworks had to be made of flexible materials.

WARNING: SPACE IS DANGEROUS

Launching a manned or unmanned spacecraft, let alone steering one around the heavens after launch, is indeed a hazardous business. Space rocketry in the 1960s and 1970s was still in its infancy, and lessons were being learned slowly and painfully, even fatally. Prior to the *Challenger* disaster of 1986, space managers tended to behave optimistically about space risks. But it was already known from surveys done on the first few tests of new rockets that, in the early 1980s, some 60 percent of them experienced at least one failure. Failures have occurred repeatedly all along the line from the development and testing, to the performance of hardware and software in virtually all launching countries, including America and Russia.

The *Challenger* event, however, was a salutory warning. The statistical risk of this kind of space tragedy began to be increased by scientists from one in 100,000 to, ultimately, one in 100. Engineers from then on began to revise their calculations. One French insurance firm recently showed that in recent years, almost two-thirds of new launches have had at least one failure. Prototype rockets, such as the Boeing Delta-3, the Ariane-5, the Lockheed Martin Atlas-3, and the Proton M from the Russian firm Khrunichev, all failed on their maiden flights. Airclaims, a London company that advises insurers about aviation industry claims, found that 12 out of 23 new designs of the 1990s suffered failures the first time around.[6] Worse, other experts suggested that the probability of at least one launch failure resulting in the loss of cargo—through an orbiting error or even an explosion—is as high as 99.5 percent.[7]

Over the years, it seems, the Russians have had more than their fair share of disasters. Since 1962 over 2,500 *Kosmos* satellites have been sent skyward with varying degrees of success. Their early attempts to get a craft close enough to capture meaningful images of Mars and record valid data or to enter orbit to relay long-term information were not, on the whole, successful. To their lasting credit, however, the Soviets were the first to land a man-made object on the surface of Mars. This was *Mars-2*—the first successful deployment of an orbiter and lander.

Still, too many mishaps occurred largely because the Soviets tried to follow on too quickly from the successful 1957 *Sputnik* launch. Excessive haste is often blamed. During the Soviet era most rocket disasters were kept secret. The first seven Soviet launches between October 1960 and November 1964 all came to nothing; four didn't even manage to leave Earth orbit, and two lost contact en route.[8] *Kosmos 57* was one of a test series for the Voskhod manned spacecraft, which broke up soon after reaching orbit. The first unmanned Soyuz spacecraft was ready for lift-off at the end of 1966, but had to be destroyed as it lurched towards China.[9] Later, in April 1967, the famed cosmonaut Vladimir Komorov was killed aboard the ill-fated *Soyuz-1*, powered by a Kosmos rocket.

Probably the worst Soviet accident happened in 1960 when Soviet leader Nikita Krushchev ordered a new rocket to be test-fired before it was ready. Dozens of scientists, including Marshal Mitrofan Nedelin, commander of Soviet Strategic Rocket Forces, were incinerated at the launch pad when the rocket blew up. It is now known that in 1980, 50 Soviet technicians were killed when a booster rocket engine exploded while being fueled. Recent faults have accumulated more noticeably since the collapse of the Russian economy in the 1990s. This has led to a rapid decline in quality control in the Proton factory, although, surprisingly, in the rapidly growing commercial satellite industry, Russia's Soyuz and Proton rockets now have a reputation for reliability.

THE SPACE STATION ERA

The global potential for further space disasters occurred at the end of the Apollo era. One of NASA's goals was to launch an ambitious series of space stations, such as the *Skylab*, by adapting for use much of the still operational *Apollo* launchers and modules. These became part of *Skylab*'s living quarters. Even parts of the Saturn rocket itself were utilized. The third stage of the Saturn rocket that was designed to hold the extra fuel to send *Apollo* to the moon was filled with air and transformed into cabins and work areas.

Nevertheless *Skylab* (officially known as the Manned Orbital Laboratory [the MOL]), launched by another powerful Saturn-5 rocket, had chalked up, within its own terms, some notable engineering and commercial achievements. The spacecraft was completed in 36 months by Mcdonnell-Douglas at a cost of $2.5 billion, before taking off on 14 May 1973. It was used by three separate crews between then and February 1974, with each crew spending a longer time in space. The various astronauts conducted intensive medical experiments and worked a sophisticated solar observatory that included x-ray, infrared, and visible light cameras, providing the most detailed observations of the sun ever made.

Yet many of the astronauts complained about poor laboratory layout and working conditions. Soon, there were fears that *Skylab* was beginning to fall apart. Unforeseen aerodynamic stresses had torn off the micrometeoroid shield and the sunshield on the spacecraft's outer skin, to drift helplessly in space. One of the two solar panels had jammed shut, and a second panel had been ripped off, crippling the craft's electrical system.

This event put back the launch of one three-man crew in the smaller Saturn 1B rocket by ten days. Their first task on docking with *Skylab* was to free up the surviving solar panel. With temperatures inside the module reaching 51°C, the astronauts—Pete Conrad, Joseph Kerwin, and Paul Weitz—had to drag a wrapped-up metal sunshield out through an airlock of the command module, which they then had to tie down to the station on a spacewalk to defend the craft—which housed an important laboratory compartment—against the sun's fierce heat.

Salyut was a thinly disguised attempt to copy *Skylab* and was built essentially for military purposes. But *Salyut* was assembled from scratch throughout the 1970s instead of using preexisting modules, although, like *Skylab*, the *Salyut* series was launched in segments. The

Russians too began to suffer problems similar to those of the Americans. The first cosmonauts to occupy the *Salyut* in 1972, Vladislav Volkov, Viktor Patsayev, and Georgi Dobrovolsky, were killed during reentry at the end of their 230-day mission.[10] But the later performances of the *Salyut* series soon improved and even began to surpass those of *Skylab*, the lifetime of which was only about a year.

In the meantime, while building the superior *Mir* from the late 1970s onward, the Russians decided to take five earlier *Salyut* space stations out of orbit as they clearly had served their largely experimental purposes. When the first *Mir* segment was launched on February 1986 by the Soviet Union, its name, which means "peace," was regarded as a promising new beginning for the space age. *Mir* was regarded as being safer and generally more technologically advanced than *Salyut*. It was launched at the time of the *Challenger* disaster and a spate of other U.S. and European rocket failures, so it was thought of as a demonstration of technological prowess in the last remaining years of Soviet power.

Mir's original life expectancy was short, as we saw in Chapter 1, although in fact it lasted 15 years, with five extra modules having been added during its lifetime. Each module seemed to grow in size, weight, and complexity, and was literally bolted on bit by bit while actually in space. The term "docking," so often used in manned space flight exercises by the Americans and Russians, was becoming a fine art. The ensemble, over time, began to look weird. Four of *Mir*'s modules had been attached at right angles to the core. Most of the segments had enormously long wing-like solar panels, making the entire craft look like a dragonfly. One American astronaut said *Mir* looked like a "cosmic tumbleweed."

By the time *Mir* was brought down in 2001, it had made 86,320 trips around the Earth, clocking 2.1 billion miles. The completed space station at the end of the 1990s had played host to a total of 23,000 scientific experiments. In fact, *Mir* had become a temporary residence for more than 1,000 humans high above the Earth, half of whom came from beyond the Russian homeland. It had provided training and research opportunities for more than 60 Russians, 7 Americans, and visitors from 10 other countries. It was still receiving crews, including Yuri Baturin, a former national security adviser to President Yeltsin, in August 1998.[11]

Furthermore, *Mir*'s expenses were dwarfed by the costs of running NASA's aging space shuttle about eight to ten times a year at $300 million for each flight—in other words, approaching $3 billion a year.

This is why one U.S. senator described the shuttle as the single greatest sink for U.S. tax dollars.

But soon the downside of *Mir* became apparent. For some years *Mir* had seemed increasingly decrepit. The longer it stayed in orbit, the more likely things were to go wrong—it might even become a runaway, out-of-control, man-made missile. It was becoming clear that its continued scientific benefits would soon be overshadowed by potential dramatic disbenefits to mankind. Roger Walker, a space debris specialist at the Defence Evaluation and Research Agency (DERA), said bringing down *Mir* was a better option than leaving it in orbit.

Problems started early: from 1987 to 2000, computer and power failures frequently occurred. Notoriously, the space station collided with a docking supply module sent up only a year after the initial core module was launched. True, *Mir*'s life was extended by Russian engineers who had added extra modules and had done virtually continuous repairs. Yet these unremitting workloads and repair schedules themselves had the Russian cosmonauts demoralized and exhausted. The craft began to reek of prolonged, rather unhygienic, human occupancy.

Although the Russians were more conscious of safety than was appreciated by their critics, even the upbeat French astronaut Jean-Pierre Haignere, who spent some six months on the space station and thought well of it, admitted that procedures were not applied strictly enough. Equipment was often obsolete. Foreign crews complained of confusing instrumentation, risky spacewalks, and bad signal audibility when communicating with ground control. Although generally dismissive of the alarmist remarks about *Mir*'s collapsing state, Haignere did suggest that socioeconomic changes in Russia had reduced the safety margins of their project.[12] He expressed surprise that the quality of work and technical ingenuity of his Russian collaborators had not deteriorated despite their not having been paid for months. Many other visiting astronauts paid tribute to the Russian cosmonauts' stamina and resourcefulness.

We saw in Chapter 1 that the year 1997 was the beginning of the end of *Mir*, with the Russian space agency taking the first tentative steps to prepare for deorbiting the craft *Mir* within a few more years. What galvanized the Russians to further this aim was the fact that 1997 was *Mir*'s worst year ever, with air conditioning problems and a near-fatal fire lasting 14 minutes. Two oxygen generators broke down. Temperatures at one stage raged out of control after corroded cooler pipes leaked ethylene glycol into the craft. Finally, the project was drumming up maintenance costs of upward of $400 million a year, which the Russians could ill afford, especially after the financial crisis that overtook the country in 1998. The disasters did not even stop there. The crew of four had a further traumatic escape from death when an American military satellite, launched by the Pentagon in 1994, nearly collided with it.[13] It passed within a mere 500 yards of the Soyuz escape capsule, where the crew had scrambled for safety.

In theory, *Mir* should have come down, literally, when the Berlin Wall and the Iron Curtain came down. The 140-ton craft had long been supported by the elite in Russia's space administration, the Council of Chief Designers, who feared the loss of 100,000 jobs when, as the Russians confirmed, *Mir* was finally closed down in March 2001.

What finally clinched *Mir*'s demise was the event that took place in June 2000. The three-man crew were attempting to execute a docking maneuver with Progress-N, a supply module as big as a double-decker bus. The radar system on Progress provided range and velocity data via a telemetry link-up. But Progress veered out of control after thruster guidance rockets were fired without the use of the TV link, which was discarded after it had malfunctioned in earlier tests. Progress thus crashed into the Spektre module of *Mir*, one of six that made up the space station.

This was said to be the worst collision in space to date. By damaging the solar panels, this accident reduced *Mir*'s power by half, and this had a bearing on how much oxygen could be generated. The solar array had been smashed, and a coin-sized puncture appeared in the side of the Spektre module, causing a slow decompression. Oxygen started to leak out of the stricken module.[14]

The Russians and their backers decided *Mir*'s time was up. Yet for a while *Mir* appeared to be granted a new lease on life. The Russian space industry was by the 1990s already heavily commercialized, and ultimately it would be completely so. Further, the ultimate aim of Daniel Golding, NASA's chief administrator, was to turn the International Space Station itself over to the privatized global space industry, including that of the Russians. *Mir* had in fact already been privatized and was an adjunct to the commercial space business: the RKK Energiya Company of Korolev, near Moscow, which partly owned *Mir* in partnership with Energiya's American division.

Things seemed to improve momentarily in the spring of 2000 after the U.S.-Canadian company Mircorp, based in Bermuda, claimed to have raised £30 million to keep *Mir* aloft for the foreseeable future.[15] This was an attempt to commercialize the project further, alongside Energiya-America, with "space tourists" in mind. It was thought that companies might wish to train astronauts or space scientists, or might want to use the main module for pharmaceutical research.

But, as we have seen, optimism was short-lived. Financial rescue plans continued for a few more months. It would be a waste and a disgrace, said Geoffrey Manber, head of MirCorp, to let *Mir* simply fall into the ocean. But the final end-game decision came in December 2000, when mission control lost contact with the craft.[16]

Mir's final demise ended its existence less than two days after the first crew of the *ISS* returned to Earth. One journalist said it was akin to a generational baton-passing to the next orbiting home. "This is the last thing *Mir* can give to mankind—the experience of how something like this can be brought down," he said.[17]

The future of the *ISS* mission, in any event, was why the Americans were pleased to see *Mir* deorbited. It was important that Russia's space heritage not be squandered. Russia had inherited excellent launch facilities from the former Soviet Union, and the Russians have an enormous amount of experience of orbital spaceflight.

CHEAPER, FASTER, MORE HAZARDOUS?

The enormous expenses involved in space missions, plus the horrendously high costs entailed in repeated malfunctions, means that the drive to reduce costs has reached almost manic proportions. Yet clearly, costs cannot be reduced without compromising the very reliability needed to reduce catastrophic failures by spacecraft and satellites, and at the same time increasing the risk of space debris.

Rocket launch companies will be virtually locked into a high-cost loop for years to come. In fact, NASA's aim of reducing the average cost per mission to as little as $50 million in 2004 now looks reckless. Some critics, like John Logsdon, director of the Space Policy Institute at George Washington University, says that penny-pinching has gone too far. French insurance brokers say that rocket designs are less reliable than they were 20 years ago, mainly due to cost cutting. NASA's views on rocket reliability have been criticized as being unrealistic. "The assumption will be that there will be no failures," said Thomas Young, a member of NASA's advisory council, consisting of engineers, scientists, and industry representatives set up to advise the space agency.[18]

NASA is now reassessing its "cheaper, faster, better" philosophy. A U.S. Atlas rocket carrying a satellite blew up shortly after launch in January 1997.[19] In the same year two Earth-observing craft, *Lewis* and *Clark*, failed in their orbits, victims of test delays and instrumental problems.[20]

In the light of these and other failures, attitudes began to change, especially after they were highlighted in a critical and admonishing NASA report. Yet things continued to go wrong as the report was being digested. An enormous financial loss involved the failure of another *Atlas-4* launch of a $1 billion Milstar intelligence satellite, which was left stranded in a useless orbit. There were three other successive *Titan-4* failures in the following two years after the report came out. A faulty Athena rocket lost a satellite. There were two Boeing Delta II mission losses.[21] A Delta-III booster failed shortly after launch from Cape Canaveral while carrying the United States Air Force (USAF) *Argos* technology satellite and South African and Danish minisatellites, and had to be detonated by a self-destruct command.[22]

An alarming incident occurred when the Solar and Heliospheric Observatory (SOHO) ran out of power after NASA, operating jointly with the European Space Agency, had lost contact with it in June 1998. The harsh conditions of space actually froze its hydrazine fuel, and its instruments seized up. Contact was regained in August when the fuel pipes thawed out, and SOHO was steered back onto course, but it was clear that some solar data had been lost for good.[23]

In another incident, 12 comsats belonging to Globalstar Telecommunications Ltd. were lost in September 1998 when a Ukrainian Zenit rocket crashed after launch from the Baikonur cosmodrome in Kazakhstan.[24] A Boeing Delta-2 was finally relaunched from Vandenberg in February 1999 after no less than 16 other botched attempts, most of which had had to be cancelled due to bad weather or technical faults. In most cases the guidance systems were thought to have overcompensated for a slight roll error on the booster during the early stages of its flight.[25] Other problems occurred in 1999. The *Terra* exploratory satellite, formerly known as *EOS AM-1*, was delayed for months after concerns about the upper stage engine caused its Atlas-2A launcher to be grounded. It finally lifted off in December 1999.

These recurring disasters, especially involving the reconnaissance and science satellites, were said to be the most expensive unmanned launches in Cape Canaveral's 50-year history, amounting to the loss of billions of dollars. A further damning report by an independent study group for NASA, published in early 2000—one of the most important reports since the shuttle disaster of 1986—repeated what had been said in its 1997 report: that cost-cutting was jeopardizing space safety.

Yet many space analysts admitted that NASA has shown an astonishing facility for snatching at least partial victories from the jaws of defeat. Damaged satellites have been brought back from the dead by their engineers, and they have made myopic space telescopes see again. *Apollo* astronauts were even rescued from assured disaster halfway to the moon. The cause was a scandalous example of careless workmanship involving a single switch that controlled an electricity supply of just 28 volts.

ROCKET FAILURES GO INTERNATIONAL

However, the string of rocket failures in the late 1990s had come to the attention of U.S. senators and sparked a Congressional investigation. The White House National Security Council (NSC) also teamed up with the Office of Science and Technology Policy to review the future of U.S. military and civil space projects. As the twentieth century came to a close, and the space industry had become a global business, a succession of embarrassing rocket mishaps took place at many of the world's launching sites. These also gave some hint as to the great number of rocket flights that now take place over considerably shortened time spans. Some analysts were saying that cost-cutting was not just an American problem—it had gone global. The NSC suggested taking a closer look at cross-investment schemes that meant that many American space projects were launched in foreign-built rockets and despatched from foreign sites on increasingly shoe-string budgets.

Apart from lack of monitoring control, subcontracting out satellite manufacturing and cross-investing in launch facilities can often create difficulties in ascertaining who is to blame when launches go awry. The Americans were by now launching satellites with Russian Proton rockets from the Baikonur launching site in Kazakhstan. Yet Proton rockets were later grounded after one of them crashed on Kazakh territory shortly after the launch in July 1999 of a Russian military comsat.[26] Two other Protons blew up seconds after lift-off in 1999, leaving the future of international space cooperation in a dangerous limbo-land until its causes were determined.[27]

Many rockets launched from China, as well as from Kazakhstan, are in fact Russian. The Chinese complained about being blamed for the failure of a Russian Proton rocket that powered the *Asiasat-3* satellite into space on Christmas Day 1997. The rocket's fourth stage failed, unleashing the satellite into an elliptical transfer orbit that took it up to the required 36,000 kilometers and then back down to within 1,000 kilometers of Earth. Only ingenious footwork by ground controllers rescued it. They fired the satellite's engines to thrust it into an upper orbit that was the highest ever known, and then gave it a final reverse kick to send it back down to change the inclination of the orbit.[28]

Some critics are now suggesting that turning space operations (including the shuttle) over to private contractors in 1996 to streamline operations may have gone too far, resulting in systemic problems that threaten safety. There was also a loss of key staff members through redundancies, and increasing workloads and stress on the remaining workers. The congressional report suggested that NASA return to employing its own workers, with more hands-on involvement.

The alternative argument in favor of subcontracting out satellite manufacture is that commercial bodies can routinely buy five, six, or even seven copies of a satellite. When one malfunctions in space, the fault can be detected and remedied in the remaining satellites. This also goes for satellites that aren't related. One built by, for example, Lockheed Martin often uses the same subsystems as those employed by, say, Hughes Electronics.

Japan is also tied in with the global space industry, but is trying at the same time to operate independently. It, too, has had its share of disasters. The Japanese always thought they were ahead in applied high-tech industries, but a string of costly and embarrassing accidents in Japan's aerospace and nuclear reactors has brought this into question.[29]

Ground controllers with the National Space Development Agency (NASDA), one of Japan's three space agencies (all due to be amalgamated into one NASA-style organization in 2003), lost contact with the Advanced Earth Observing Satellite (ADEOS) on 30 June 1997 after its solar panels stopped functioning. Vital climate data were irretrievably lost, including a radar scatterometer built by NASA that measured sea surface winds.[30] It had a troubled start, having to use a back-up propulsion system to reach its planned orbit in 1996. At first, officials thought it had been hit by space junk.

Later in November 1999, an H2 heavy-lifting rocket carrying a government satellite costing $340 million crashed into the sea shortly after take-off. It was blown up by command from ground controllers minutes after its launch, when the main engine cut out.[31] NASDA had an order for ten launches from Hughes Electronics cancelled in May 2000, following repeated failures of its H2 rocket.[32] This was seen as a body-blow to Japan's ambitions as a commercial satellite launcher. It caused the top bureaucrat at the Ministry of Science and Technology, Toshio Okazaki, to resign his post as deputy minister.

Just three months after this event the three-stage M5 rocket, carrying a joint NASDA-NASA x-ray astronomical device called Astro-E and costing £105 million, was launched from the Kagoshima Space Center in Uchinoura in southern Japan. But it too suffered a massive failure. Less than a minute after take-off, the first stage of the rocket did not fire properly, and the M5 released the satellite at a suborbital height, causing it to fall back to Earth and burn up in the atmosphere. Some reports said the craft spun out of control after graphite on the rocket's nozzle appeared to fall off, exposing it to heat damage.[33] No information was given in the scientific press as to whether it burned up entirely (which seems unlikely) or whether it landed in the sea or on land.[34] Later, the replacement—a slimmed-down H-2A 290-ton 172-foot-long (52 meters) rocket—was finally successfully launched in August 2001 after three launches in a row were aborted.[35]

Why was Japan also experiencing space launch failures? Japanese research and development is as strong as ever, but its applied innovation increasingly leaves much to be desired. Hiroaki Yanagida, an engineering professor at Tokyo University, blamed complacency on the part of the engineers. Researchers at the Massachusetts Institute of Technology think much of the malaise is simply psychological. Possibly there are flaws in technology training, or perhaps technicians prefer to work in consumer electronics, or students increasingly aim for academic training instead.[36]

However, if the longer view is taken, it must be said that steady progress to successful launch capabilities has been the hallmark of both South and East Asia. Much of this stems from a desire to operate an independent spacefaring policy untrammelled by complicated multilateral interdependence, which inevitably requires tolerance of frustrating malfunctions of rockets owned by other organizations.

Both Koreas, North and South, may have the same lofty goals in sight: joining the world's spacefaring nations. North Korea claims that its satellite, called *Kwangmyongson-1* ("Little Star") was successfully placed into orbit and had completed 100 orbits by 14 September

1998. But no satellite was traced and no transmissions were heard by U.S. Space Command, and there were fears that it had crash-landed somewhere. North Korea's ambitions to join the world's space station missions were halted in August 1998 when an apparent third-stage booster failed.

What was troubling to western defense experts was that the launch rocket used was also a two-stage ballistic missile, which flew over the Japanese mainland and caused an international incident. The missile was a Taepodong 1—based on the Soviet Scud. It was feared that the North Koreans were near to developing the Taepodong 2—with a range of 6,000 kilometers, which could carry a warhead into the western seaboard of the United States.

Now South Korea plans to free itself from its dependence on other countries to launch its satellites. But western security experts have expressed concern that the launching technology could also be sold to other countries to make long-range ballistic missiles if the satellite project encountered difficulties or was loss-making.[37]

Whatever the consequences, the sale of launching technologies, whether or not for the purpose of sending guided missiles to attack another country, could only result in more man-made missiles arriving from outer space. This aspect will be the focus of our next chapter.

CHAPTER 3

Man-made Missiles from Space

On New Year's Eve 1978 in northern Europe, thousands of witnesses claimed they saw a trail of lights in the sky. People reported seeing the lighted windows from a "cigar-shaped craft." British Ministry of Defence investigators and space scientists called in to investigate the matter concluded, however, that this event was simply infalling space debris that had taken on a rather surrealistic appearance. In fact, it turned out to be a spectacular reentry into the atmosphere of a *Kosmos* satellite, launched in Russia only a few days before. The metal fragments were heated by friction, forming a chain of debris miles long in the sky. The "windows" in the "cigar-like craft" were optical illusions arising from the persistence of vision, rather like "joining the dots" in a puzzle.

Clearly some debris may be cosmic in origin. Space debris, passing through Earth's atmosphere, can take in excess of 30 seconds to cross the sky and is much slower than incoming meteorites. Many of the blazing trails of lights they create when doing this have often been attributed to meteorites. If the event takes place within 100 miles of an observer, the debris can even appear to be rising.[1]

The normal consequence of reentry of any solid object larger than a particle poses considerable dangers. Heavy metal fragments, when they do not land intact, are unlike natural substances like dust or ice and can produce vivid colors as they break into molten blobs. In 1997 a pilot reported a fireball over Chicago that streaked across the sky before going into a steep descent. It lit up a cloud deck as it passed through. Analysts at the North American Air Defense Department (NORAD) believed it to be a piece of decaying rocket debris, and it was cataloged as item 24245, part of a Step-2 Pegasus vehicle launched in 1994.[2] During the nights of 30 and 31 March 1993, bright lights were seen over many parts of northern Europe and southern Britain, accompanied by vapor trails. This was also alleged by NORAD to be the burn-up of a Soviet military satellite, *Kosmos 2238*.[3] A huge cloud of particles that hung high in the stratosphere over San Francisco Bay for several days in April 1997 was explained by NASA's Goddard Space Flight Center as the product of a kerosene plume from a Russian rocket launched from Baikonur in Kazakhstan only two weeks earlier.[4]

Scientific knowledge about the way different types of space debris behave is invaluable. It provides an important yardstick for determining whether glowing objects that appear to have a fast trajectory are indeed cosmic objects. If the incoming projectile is also accompanied by a loud explosion, then it is more than likely to be some sort of bolide (a cometary or asteroidal object).

One Sunday in March 2000, local people living on or near Easter Island in the Pacific saw a brilliant streak of light in the skies as a Proton rocket blasted skyward from a converted oil platform. Several minutes later they saw another flash of light followed by a muffled explosion as the same rocket crashed into the sea, destroying its multimillion dollar communications satellite system.[5] The satellite was the first in a series built by Hughes Electronics for ICO Global Communications, a consortium striving to gain the competitive advantage over its rivals by building the "ultimate" global mobile phone network.

A few months later, a 17-ton space observation satellite the size of a railway engine came crashing back to Earth after it was destroyed by ultracautious NASA scientists. This was the Compton Gamma Ray Observatory (CGRO), which had orbited the Earth 51,658 times since 1991 and would normally have survived until 2005. (The streaks of light seen from Texas to Nebraska on 1 December 2001 were also said by the North American Air Defense Agency [NORAD], in a report to both the National Weather Service and the Federal Aviation Authority [FAA] at Minneapolis, to be space debris.)[6]

The CGRO had in fact worked well until a gyroscope failed. This itself would not necessarily have caused difficulties in keeping the spacecraft on its correct trajectory, but if another had failed, it might well have crashed onto land without burning up (because of its size and low orbit), thus endangering human life. Edward Weiler, associate administrator of space science at NASA, put the chances of a human fatality if the CGRO went out of control at one in 1,000: still a big risk.[7] This had the potential of becoming a scientific, financial, and public relations disaster for NASA, whose recent missions have been beset by technical blunders and managerial ineptitude.

Some scientists at the NASA Goddard Space Flight Center in Greenbelt, Maryland, such as Neil Gehrels, criticized the decision to deorbit the CGRO, saying that alternative technology could have kept the satellite on course for several more years without having to deorbit it, and a controlled descent could still be done without any gyros.[8] But having the ability to bring the satellite down where they wanted, the scientists, in the first known planned and controlled crash of an expensive observatory, aimed for the familiar target area in the Pacific, this time some 2,500 miles southeast of Hawaii and at least 600 miles from the nearest land mass. NASA fired six burns of the Compton's thrusters to lower it into the atmosphere. In fact, the controlled descent of the CGRO was a triumph for space scientists, as was the deorbiting of *Mir* a year later. In earlier years the fate of failing spacecraft could not have been handled in this manner by either American, Russian, or Japanese space scientists.

These observations of "lights in the sky" by ordinary eyewitnesses are nevertheless significant. In the last chapter we discussed the hazards of constructing space stations in space and the various mishaps that have occurred in the commercial launch industry. Space industry spokesmen point out that as launch sites are invariably near the coast (for example at Cape Canaveral at a North Atlantic site or at Kourou in French New Guinea adjacent to the South Atlantic), the malfunctioning rockets fall within minutes into the sea.

This is the theory. *Salyut-6*, weighing 40 tons, was successfully brought down into the Pacific in July 1982. However when *Mir*'s successor, *Salyut 7*, launched on 19 August 1982, came down on 7 February 1991, Russian ground controllers ran low on fuel and tried to steer the 40-ton craft into the Atlantic, but it was overpowered by aerodynamic effects. Large chunks of it plunged instead into Argentinian forests, starting several fires. One piece is said to have plunged into the garden of a house where a woman was doing her ironing.

One of the last great Soviet space projects, the *Polyus* prototype weaponized space station, was the Russian response to "Star Wars." The 37-meter-long cylinder was painted black and armed with a laser weapon, nuclear mines, and an antiaircraft machine gun. It was launched in May 1987, but a thruster malfunction sent it plunging back to Earth. However, much of its 80-ton mass would have survived the fiery fall. Russian and western intelligence agencies believe it lies somewhere under the South Pacific,[9] but no one can be sure.

In the meantime, what happened to *Skylab*, already "falling apart" as some described it? *Skylab* was eventually deorbited in July 1979, having drifted aimlessly for five years. NASA engineers had limited control over the 75-ton station. To improve its degree of control, *Skylab* would have needed an additional fuel cargo bay and accompanying thrusters, which would have increased its payload. As we shall see in a later chapter, space conditions may also have interfered with its trajectory. Instead, controllers changed *Skylab*'s orientation in space to increase or reduce atmospheric drag in order to enable it to home in on a chosen entry point. It scattered burning debris above the Nullarbor Plain in western Australia, with some of the debris landing near a remote sheep station close to the small town of Balladonia, 850 kilometer east of Perth.[10] One piece of it fell on a golf course in Albany, western Australia. It measured one meter, and weighed one kilogram. Other scientists found a water tank, an airlock shroud, oxygen bottles, and a film vault, all scattered along the deorbiting track. The bulk of *Skylab* fell into the Indian Ocean.[11] NASA vowed to prevent further uncontrolled descents of such massive spacecraft.[12]

China is now a big player in the launch business. It has used Russian rockets, as we have seen, but increasingly uses its own. But a series of rocket disasters in the mid-1990s led to China's losing a contract with the Intelsat organization.[13] Its launch program led to worldwide anxieties as rockets and satellites veered erratically through Asian skies. In one notorious incident, two of China's Long March rockets blew up near the ground, killing more than 120 civilians.[14]

In another incident, the Long March 2C booster appeared to have been successfully launched, to much cheering and applause, in October 1993 from the Quiquan Center in the Gobi Desert. On top was lodged their one-ton *FSW-1* satellite. It was due to spend just a few days photographing Earth from space, after which it should have jettisoned a module containing the monitoring equipment. But things went disastrously and expensively wrong when ground control errors activated its rockets at an inauspicious moment. The reentry module was sent spinning into an unstable elliptical orbit to swing around Earth every 100 minutes, dipping into the upper atmosphere at its closest approach at 100 miles above the Earth.[15]

The Japanese, hoping to learn a thing or two to their advantage and improve their own launch record, showed close interest in the career of *FSW-1*. So did British space watchers. So did the western media, for that matter. Not normally given to publicizing space news, the careening satellite got more attention than the *Challenger* disaster. This was because the reentry in the spring of 1996 was originally believed to bring the craft crashing down on any populated part of the southern hemisphere. Unlike most satellites, it would not burn up on reentry because it had a heat shield designed to withstand 1,200°C. It was all the more surprising that the *FSW-1* could withstand such temperatures, heatshield or no, because it was revealed that parts of the main structure of the craft were actually made of the most combustible of materials: oak wood!

But it could still be travelling at well over 1,000 mph as it approached the surface. "It would cause devastation if it landed in a built-up area," said Professor Alan Johnston of the Mullard Space Science Laboratory at University College, London. "They [the Chinese] do not know where it is going to land and they cannot do anything to regain control." It was kept under close surveillance by DERA, an arm of the Ministry of Defence at Farnborough, Hants. Ultimately the *FSW-1* fell into the Pacific.

As we saw in Chapter 2, space disasters seemed to accelerate as the last decade of the millennium came to an end. In 1997 an Athena rocket took off from Vandenberg in California but crashed into the Pacific with its *Ikonos-1* commercial spy satellite. A Lockheed Martin Titan-4 launcher exploded over the Atlantic in 1998, destroying a reconnaissance satellite.[16] An attempt, also by the United States Air Force, to launch a spy satellite a year later failed to reach its proper orbit when the upper stage of the rocket malfunctioned.[17] A Titan-4 rocket again did not successfully place a military comsat into its intended Geosynchronous Earth Orbit (GEO) the following year, in May 1999, and it exploded over the Atlantic. Two weeks after this, the second stage of a Delta rocket shut down after only one minute's burn, leaving a comsat in an equally useless Low Earth Orbit (LEO) about 150 miles up in the lower atmosphere. The National Oceanic and Atmospheric Association announced that it was delaying further weather satellites until a cause of the repeated rocket failures was diagnosed.

Other spacecraft likely to re-enter in an uncontrolled manner are *Abrixas*, an x-ray satellite with a mass of 500 kilograms that malfunctioned shortly after its launch in April 1999; *Sax*, a similiar science craft; and *Rosat*, which has a mass of 2.4 tons and a massive x-ray mirror made from high-melting glass (Zerodur) on board, but is still regarded as likely to survive re-entry. *Rosat* was said to be moving uncontrollably since the end of its mission in February 1999.[18] Robin O'Hara of Lockheed Martin Space Operations in Houston said that the scientific Ultraviolet Explorer science satellite (*EUVE*), launched in June 1992, could de-orbit out of control with the risk of human casualties put at 1 in 5,300. This is because the *EUVE* spacecraft was not designed with a propulsion system and therefore could not have a guided re-entry, which requires ground controllers to fire course-adjusting jets.[19]

EUROPEAN SPACE MISADVENTURES

Things were not much better with European launches. Ariane-4 had a patchy record, and the European space consortium Eutelsat flew its *W2* satellite on it. The follow-up to Ariane-4, the Ariane-5, was viewed as a successful potential participant in the growing world-wide comsat industry following repeated disappointments with the earlier Ariane rockets. Understandably engineers took their time preparing the successor Ariane-5 rocket. Another flop would wreck the image of European rocketry.

Yet repeated failures of even the new Ariane-5 were viewed with disappointment and even alarm. In all it had taken ten years and £6 billion to develop the Ariane-5 series, designed to place 18-ton payloads in space, enough for a manned spaceship. The maiden flight was to have been made in 1995. But the big booster did not lift off until June 1996 and promptly broke apart 30 seconds later.[20] This disaster, it turned out, was due to a computer having been programmed to expect an Ariane-4 type launch, thus causing the later variant of booster to veer off course. Four other expensive ESA science satellites were lost in similar circumstances.

Other lessons were learned by the French in regard to malfunctioning satellites. Although every part is thoroughly tested, and despite advances in propulsion technology and the increasing use of computerized flight-control guidance systems, faults inevitably occur. Things can go wrong because a satellite might not be properly released from its fairing. It could end up in the wrong orbit, or its electronics could go wrong. Engineers design and build instruments sometimes to tolerances finer than a micron (a millionth of a meter), then place them on top of a rocket containing vast quantities of highly inflammable fuel. The onboard instruments are then subject to a bone-shaking ride through Earth's atmosphere. The rocket of one of the stages could well fail during lift-off or when heading towards orbit since failures at the booster stage in near space, as we saw in the last chapter, are more common than explosions on the launch pad. Titan Transtages have been known to exhibit abrupt orbital changes, possibly due to malfunctions.[21] All the while the craft must operate on less electricity than would power a 40-watt bulb. Once in space there is no chance to do repairs, even if it were thought to be financially worthwhile to do so (the exception, of course, has been with the manned spacecraft like the *Mir* or *ISS*, or prestigious pieces of equipment like the Hubble telescope).

Says David Todd, who edits the SpaceTrak database, "A maiden flight is a brand-new vehicle—it's not like an aircraft that's been test-flown in small altitude and speed steps. With a rocket all your eggs are in that one flight, and there's always a chance of failure."[22] In other words, the craft's design must be gotten right at the very start. Hence the margin for operational mistakes in building and programming rockets is tiny. "Space people are conservative. Nobody wants to be the guy who tried something that failed," says Marc Rayman, a NASA engineer.[23] "People must understand how difficult these (space missions) are. It's like a contact sport and your playing field is the whole solar system. It's full of hazards and unknowns," said Wayne Zimmerman, a head electronics engineer at NASA.[24]

If there is a need for quality control on the assembly line, then the problem is clearly one of costs—and the law of diminishing returns arises. In other words, the assured safety benefits do not accrue commensurately for every dollar spent. According to Jay Chabrow, leader of JMR Associates, a technology consulting firm in Las Vegas, launch vehicle failures are more or less inevitable no matter how good the quality control. Repeated launches must raise the likelihood of failure. In fact, of 12 launches a year, calculated as being 0.92 to the power of 12, one would achieve a 0.386, or 38.6 percent, success rate.

The French decided to proceed cautiously with the new Ariane-5. A further attempt at a fault-free launch took place in October 1997.

The flight was meant to demonstrate the ability to glide successfully from geostationary transfer orbit (GTO) to the final geostationary orbit (GEO) some 36,000 kilometers above the the Earth's surface. Yet Ariane-5 experienced an unfortunate "roll torque," a kind of rotation, that failed to keep the rocket perfectly on course. This in itself was not disastrous. But onboard satellite fuel in Ariane's booster rockets has to be used sparingly. The fuel has to last several years and must not take up too much payload space. The roll torque problem meant that too much of its precious fuel would have had to be used, thus reducing its operational lifetime.

The performance of a later Ariane-5 in the same year yielded only a marginal improvement. The flight data revealed that there was a shortfall in the launcher's thrust, which left its dummy payload thousands of miles below the intended orbit. The first stage, with its solid rocket booster, came down in the western Pacific instead of off the coast of Colombia.

Again there was a fuel problem. Excessive spin—similar to the earlier "roll torque"—on the rocket around its long axis, probably caused by exhaust from turbines, forced the fuel in the first stage to flatten against the walls, causing its sensors to assume that the fuel was running low. There was an automatic switch-off of the engines to prevent the booster from exploding, and it was ditched into space. Hence this early jettisoning of the booster meant that the spacecraft was short of speed for Earth orbit. In addition, parachute failures meant that the booster was destroyed on impact, thus preventing the Centre National d'Etudes Spatiales (CNES) (French) engineers from examining the casings to see whether they had been burned through, so to enable them to make modifications for future flights.[25]

Things were a lot better a year later in 1998 after the engineers on a new Ariane-5 booster, the third of its type, had corrected the veering fault. The latest Ariane-5 included an Atmospheric Reentry Demonstrator, the last relic of Europe's manned space flight program. Even here, though, there was a glitch involving the loss of one of the two booster rockets, which broke up before the start of the recovery sequence that fires the chutes.[26] Yet the overall success of the launch of the spacecraft in November, carrying a dummy satellite, meant that the French aerospace industry could breathe easy. Otherwise it would have been a disaster for the European Space Agency and a major blow to its commercial hopes. One more failure, one more fuel glitch, would have been a severe blow for the commercial

hopes of Arianespace, from which it would have taken years to recover.

THE RISE AND STALL OF THE GLOBAL COMSAT BUSINESS

At this point we should pause to ask why there seems to be so
much determination on the part of the French, Chinese, Russians,
Americans, and indeed most major powers, to get into the increasingly hazardous launch business. Simply, it is because of the importance of the growing commercial satellite industry to their national
economies. The rise of the industry in recent years has been phe-
nomenal. According to the Organization of Economic Cooperation
and Development (OECD), telecommunications has become the
third largest sector of the global economy after healthcare and bank-
ing. The industry, which includes broadcasting, is now worth more
than 1 trillion a year among the world's 20 wealthiest nations.

We should bear in mind that about 80 percent of the world's population has no access to reliable telecoms, and about one-third live in
dwellings that do not have a fixed electricity supply. Some 40 million
people in less well-off countries are presently waiting for fixed-line
phones. In Russia the average wait is three years, and waiting lists for
many living in other countries simply do not exist. It is virtually impossible to provide phones to inhabitants of remote and isolated
Third World areas because it makes no economic sense to lay fixed
lines there. Bhutan, for example, has fewer than two phones lines for
every 100 people.

Hence mobile phones are of enormous intrinsic value to these vast,
unconnected communities. Further, cellular phone companies have
much lower costs simply because they do not have to dig holes in the
ground or lay expensive copper wire to their customers. All they need
to do is to put up masts, which can be quickly replaced even if destroyed in natural disasters.

The pressure on governments to help facilitate mobile telecommunications and satellite TV broadcasts is driven partly by the structural economic reforms that most governments are now carrying out,
mainly benefitting the largely young populations in the developing
world. The United States has the world's largest cellular phone market, with almost 80 million subscribers, but proportionally it has less
mobiles per head of population than Japan, where 37 percent of the

population owns one. In eight countries, almost all in the northern hemisphere, more than a third of the population now own mobile phones: among Scandinavian men in their twenties, the figure is nearly 100 percent. The burgeoning communications industry is now worth an added $50 billion a year to the GDP of those countries with a thriving high-tech sector.

Some observers have begun to worry about the problems that were bound to arise sooner or later if the rate of satellite organizations continues to proliferate. The end of superpower rivalry in space implies that state-run and financed civilian space programs are largely a thing of the past. Encouraging signs of the post–Cold War "swords into ploughshares" syndrome are evident. Privatization has facilitated takeover consolidations and cross-investment across and between nations, as we have seen in the case of the *Mir* and International Space Station.

New start-ups and stock market media and high-tech flotations used to be announced in the press on an almost weekly basis. European phone companies spent some $100 billion in 2000 alone on purchasing radio licenses from the authorities for their telecommunication services. Many were and are desperately trying to link up with multimedia consortia to include satellite-beamed all-in-one Internet and broadcasting facilities.[27] It was no longer clear who actually owned the satellites, or which country had authorized their launches, or which "Internet servers" operated with links via which phone company domiciled in which country. It was true globalization with a vengeance.

One problem for the licencees was the increasing shortages and overcrowding of radio bandwidths, and, indeed, spacecraft flying too close together in the crowded LEO (Low Earth Orbit) zones. One problem identified by Manfredo Porfilio at the Rome School of Engineering and Aerospace was that with lifting, for example, five satellites into similar 650-kilometer-high orbits, the spent upper stages and the satellites themselves are insufficiently spaced out, and "as a consequence potentially dangerous close approaches may occur."[28] According to the MIT Lincoln Laboratory, a *SATMEX* satellite failed in late August 2000 while in GEO (Geosynchronous Earth Orbit). With a longitude drift of between 101 degrees and 109 degrees west, it threatened to overlap the path of *Telstar 401*. There was also "significant concern" when *Galaxy 7*, owned by Panamsat, failed in its 125-degree orbit in late November 2000.[29]

The International Telecommunications Union, publisher of the

Radio Regulations (in effect an international treaty), coordinates frequency and longitude zones for GEO craft. Operators perform longitude control maneuvers, termed east-west station-keeping maneuvers, every two weeks or so.

Further, the sheer volume of satellite launches suggests that the success rate itself actually goes down stealthily, or the number of mishaps rises, although, to checkmate this fear, there are increasing bankruptices now among broadband carriers and telecom companies. Nevertheless, never has the expression "what goes up must eventually come down" been so apposite. So far there is no evidence of an actual decline in the number of satellite launches, as many are sent up to replace aging versions (see Chapter 6). Instead many satellites have been pulled out of their orbits sooner than planned.

This is because by the turn of the new century, the cumulative effects of overcapitalization in the high-tech communications sector, coupled with the crisis in the European telecom industry, meant that many companies were failing to make actual profits. By the end of the year 2001 the contraction in the U.S. economy in the third quarter reflected a sharp pullback in consumer spending and had particularly affected the multimedia industry (although there was a partial recovery in 2002). In the United States, after the September 11 attacks, the Defense Department was becoming increasingly reluctant to sell off airwaves to multimedia companies for security reasons.

But even before the September 11 attacks, the International Communications Organization (ICO)—whose first satellite, as we saw, plunged into the sea in 2000—was on the verge of going bust, having filed for Chapter 11 bankruptcy in Delaware. It had run up development costs of more than $3 billion. One of ICO's rivals was the *Iridium* series of satellites, which was also on the verge of going bust. Motorola, the lead investor, has written off more than a billion dollars on the abortive project because too few people were using the service. As Motorola was one of the first in the cell phone business, its handsets, dating back to the 1980s, tended to be rather heavy and chunky, reminiscent of the earlier walkie-talkie sets. The company soon found it was unable to compete with the new generation of cheaper, lighter, mobile cell phones.

In the early summer of 2000, the Iridium consortium, with its network controlled by the Satellite Network Operations Center in Landsdowne, Virginia, planned to send the first of its giant satellites spiralling toward Earth to burn up in what a *New York Times* journalist described as a "fitting and fiery end to one of the colossal corporate failures of recent years."[30] Five of its 72 satellites had already failed[31] before the decision was made, but later rescinded, to deorbit the entire fleet. The satellites would probably have been brought down four at a time over a period of about two years, firing their thrusters to drop them into a destructive orbit. George Levin, director of the Aeronautics and Space Engineering Board at the U.S. National Research Council, applauded Iridium for building in deorbiting facilities, with enough fuel for controllers to steer the satellites into uninhabited areas.[32]

But a spokesman for the Space Command Center, Major Perry Nouis, said that it was likely that some parts of the satellites would survive because of their enormous size.[33] The entire deorbiting project would present problems because fuel glitches or computer failures could lead to some of the intact debris falling onto dry land. The Iridium probes were and are receiving and retransmitting signals from about 20 other larger probes in deep geostationary orbit, some 22,000 miles up.

There was some dispute in the press as to whether defunct satellites could cause harm to other satellites or to aircraft or even to people on the ground if the inevitable decision is made to bring them down. According to David Mangus and colleagues at the NASA Goddard Space Flight Center in Greenbelt, Maryland, the risk to life and property from infalling debris is 1 in 1,000 if spacecraft are uncontrolled.[34] But this threat could be dramatically reduced to 1 in 29 million with careful design and guided re-entry. Some argue that the space experts know what they are doing, and by now are highly accomplished (as we have seen with *Mir* and the Compton Gamma Ray Observatory) at bringing down deorbiting satellites with pinpoint accuracy. In terms of re-entry, international sovereignty claims and NASA Safety Standard 1740 dictate that the target area should be at least 200 nautical miles from foreign lands and 25 nautical miles away from the U.S. mainland; this often necessitates the Pacific Ocean splashdown.[35] The U.S. Space Command (or USSPACECOM, of which more will be said in Chapter 7), will offer guidance to the new Iridium LLC organization about where and when to pull the satellites out of their 485-mile-high orbits, should this be necessary.

CHAPTER 4

Toxic Rain from Space

We referred in Chapter 2 to the failures of NASA's Mars missions in the late 1990s, with spacecraft crash-landing on Mars' surface. However, these failures spelled no danger for humankind here on Earth.

But not so with the Russian Mars ventures, where high stakes were often involved in just getting spacecraft into orbit. For example, their 7-ton Mars-96 probe, launched from Baikonur on 16 November 1996, was so big that it required a huge three-stage Proton rocket, as well as a kick from a fourth-stage thrust, to get into a parking orbit 160 kilometers above the Earth. It still needed a further small engine to complete the escape maneuver. Yet if it had been successful, it would have solved many of the Red Planet's mysteries and would have scored a considerable lead over the Americans for the first time since the 1960s.

But soon events were to turn awry. In theory, the propellant problem should hardly figure in the space debris scenario. Propellants are needed largely for the launch stage itself. The satellite is put into orbit by first attaching it to the last segment of a multistage rocket and then accelerating it through the atmosphere so that it reaches speeds of 17,000 mph—the so-called escape velocity. As soon as the launcher has reached a predetermined altitude and trajectory, a radio command from ground base releases the satellite. Once in orbit the machine needs little extra power. It will behave just like a minimoon because the absence of an atmosphere beyond the stratosphere, some 20 miles up, means that there will be no friction to slow it down.

In fact satellites, or "probes" will spend 99 percent of their journey in unpowered flight. Designers know that, once in space, many external influences act upon satellites. As the southern hemisphere is larger, Earth's center of mass is south of the equator. This tends to pull the probe out of its orbit around the equator. The moon's gravity also has an influence, first pushing and then pulling. Hence correcting thrusters are fitted at the corners, but give off only tiny amounts of thrust. To twist the craft, one set of thrusters fires a short burst. This keeps it rotating until counterbalanced by a burst from an opposing set of thrusters.[1]

Onboard reaction wheels that are like flywheels or speed governors are often deployed, one on each axis. Sensors detect the craft's movement and send signals to the individual flywheels to control the craft's speed. But if the thrusters are wrongly directed, the spacecraft will end up millions of miles off course. Satellite manufacturers guarantee that their machines will stay in position to an accuracy of 0.1 degrees. But even this can give rise to error: over long periods of time in space, the errors naturally accumulate.

Yet the failures of Russia's *Mars-96* probe had little to do with the guidance of the satellite perched on top of the Proton but with the misbehavior of the Proton itself. The second of the fourth stage burn was supposed to take place over the coast of Uruguay. This extra thrust, however, never went according to plan, and the bulk of *Mars-96* was left only in a parking orbit. Sensing something was wrong, the on-board computer turned the small engine off after crossing the West African coast of Cote d'Ivoire. This automatic action need not have taken place if the Russian mission controllers, having sold off most of their tracking ships for scrap, had not lost track of the craft and had monitored the failed burn stage over West Africa.[2]

Unfortunately, even the on-board computer was faulty, and confused the Russians at their one good land-based tracking facility at Yevpatoriya in the Crimea by broadcasting that the "spacecraft was in an interplanetary cruise configuration." When the truth was realized, a desperate attempt was made to deorbit the spacecraft. A full-blown emergency was put in place when it was thought that the satellite would reenter and crash into Australia. In the end, it burned up somewhere west of Chile, near Easter Island.

RADIOACTIVE FUELS

But it was the knowledge that Mars-96 was powered by 200 grams of plutonium, a hot radioactive substance needed to provide heat on the cold Martian surface, that was the cause of concern. Each of its batteries was the size of a 35-millimeter film cannister and contained about 12 grams of plutonium-238. The Americans, who were also monitoring the craft's trajectory, believed that the plutonium was still attached to the rocket. In fact it had already detached itself, but few at U.S. Space Command knew where in its orbit it had done so or where it and the spacecraft were likely to land on Earth. It was assumed to be in the mid-Atlantic.[3] "We were aware of a number of eyewitness accounts of the re-entry event via the media," said Major Stephen Boylan, chief of the media division at the U.S. Space Command in Colorado Springs. "Upon further analysis, we believe it is reasonable that *Mars-6* fell onto dry land."[4]

Despite warnings from Space Command to Chile and Bolivia, there was no standby or official help. It was a clear night, and many eyewitnesses had seen the probe descend: it was described as a bright slow-moving object from which sparkling fragments were seen to fall. The Chilean government, however, told the people that they had seen nothing important.

It is essential, of course, that rocket failures are kept to a minimum or are contained within the launch zone. The use of plutonium or other highly toxic fuels to power spacecraft has always been controversial. In fact, at least 50 existing satellites pose a threat, as many contain other radioactive materials such as beryllium and its oxides, lithium hydride, and uranium, either in nuclear reactors or in radioisotope thermoelectric generators.[5] According to Dr. Helen Caldicott of Physicians for Social Responsibility: "Plutonium is so toxic that less than one-millionths of a gram is a carcinogenic dose. . . . One pound, if uniformly distributed, could hypothetically induce lung cancer in every person on Earth."[6]

Satellites with radar need powerful energy supplies to keep them going. Starting in the early 1970s, the Soviets used uranium-235 in their reconnaissance satellites. *Kosmos 198* was Russia's Rorsat (Radar Ocean Reconnaissance Satellite), sent aloft to monitor U.S. naval fleet activity. The Rorsat had a monitoring life of little more than two months. The plan was to separate the nuclear core from the craft and boost it into a high orbit where it could stay more or less permanently. *Kosmos 954* was also a Rorsat, launched in September 1977.

Shortly after this launch, the Rorsat program shuddered to a temporary halt when the boosting procedure on the *Kosmos 954* failed. On 24 January 1978, the reactor reentered the Earth's atmosphere and scattered radioactive debris across Canada's Northwest Territories, blazing a trail across the tundra. Hundreds of witnesses across Europe also saw it and described it as a "blazing rocket-shaped device," an event that was caught on film. The glowing debris over Canada was mistaken for a UFO, in the same way that, barely a year later, the returning *Kosmos* satellite was when seen over European skies. Most of the uranium was recovered by the CIA, and the area cost the Americans and the Canadians some $13 million to clear up.[7]

The Soviets delayed further nuclear-powered flights for two years after this before launching another Rorsat—*Kosmos 1176*. Two years on and *Kosmos 1402* failed in the same way as had *954*—the nuclear reactor could not be properly boosted. On 23 January 1983, one part reentered the atmosphere over the Indian Ocean, and the following month the nuclear core landed in the South Atlantic. Russian scientists point out that there are 31 Russian nuclear-powered spacecraft (NPS) in orbit approaching 1,000 kilometers in altitude. Twenty-nine of them have thermoelectric or thermionic fuel elements (TFEs), and two NPSs have thermionic conversion systems. Three of them, *Kosmos 1670*, *Kosmos 1677*, and *Kosmos 1900*, could be particularly hazardous because "imperfect core separation from the reactor could not be ruled out."[8]

If toxic material or atomic fuel is lost over sea, no one is greatly worried. For this reason most of the spacefaring nations, such as the Americans, French, Chinese, and Japanese, have their launch sites adjacent to the sea as we have seen. The Russians, however, using the Central Asian Baikonur spaceport, do not have this opportunity, but try to steer any aborting rocket into either the Pacific or the Atlantic. Baikonur is deep inland in the Kazakhstan steppes and is surrounded by hundreds of square miles of virtually uninhabited territory.

The arctic tundra around Plesetsk in Kazakhstan is littered with space debris. Although miles away from Kazakhstan, twisted chunks of Proton missile metal can be found strewn in the wooded pine forests and high peaks of the Altai Republic. Proton rockets were grounded at the Baikonur cosmodrome after one rocket crashed near the launch pad shortly after take-off in July 1999.[9] Nevertheless, the grounding was only for a short duration, as the Kazakhs felt they had to comply with a request of the Russian space agency to fly an urgent mission to space station *Mir*. Even so, the accident came at a tense time for Russian-Kazakh relations because the Russians had leased the base and were behind in their payments. The situation was not helped when a Proton rocket crashed again in October 1999.[10]

Toxic fuels, of course, could contaminate large areas should the failed spacecraft be destroyed or damaged on impact with the ground. In several instances, out-of-control Russian craft have allowed toxic fuel to escape. On one occasion, a piece of satellite debris smashed into the courtyard of a house, and toxic fuel from the rocket leaked across a large area of central Kazakhstan, starting a fire that spread over a large area.[11] Problems in regard to discarded fuel also plague China's space project. In the western Xinjiang province this seems to be a major issue.

Alan Johnston, of Mullard in London, estimates that more than 1,400 tons of rocket parts and probably as much fuel have been dumped on the Altai Republic in Siberia over recent years. Rocket parts have also fallen over populated areas. Residents from Dzhez-kazgan, which lies in the flight path of rockets, have complained of toxic fuel from crashed rocket stages contaminating the soil while creating clouds of poisonous gases. Some 10 percent of the fuel, usually dimethyl hydrazine (or "heptyl") from a Russian rocket, remains unburned. Most of it falls back to Earth when the rocket casts off its second stage.[12] This occurs at 100 kilometers up and when the trajectory starts to flatten in preparation for satellite launching.

Heptyl poisoning also occurred when the Russian army complied with the Soviet-American Strategic Arms Reduction Talks (START) disarmament treaty and removed warheads still containing the highly toxic rocket fuel. A report by the Institute of Biophysics at Russia's Ministry of Health recorded that in certain regions of Altai, heptyl can be detected in soil and plants, with concentrations of chemicals reaching 0.3 milligrams per kilogram—considered to be a high dose. At an international conference in Minsk, Belarus, in 1993, chemists warned that the risks from heptyl had been underestimated.[13] Health problems have developed in residents in the Sakha region of Siberia; in particular, liver deformations have occurred in babies.

Alan Johnston has advised scientists at the University of Omsk, 1,000 kilometers to the northwest of Altai, on how to power spacecraft using less-polluting rocket fuel. But the Russians are years away from using such fuel, he says.

In fact further fears about toxic spacecraft fuel occurred in December 2000, when six Russian satellites were lost as a rocket booster crashed over the Arctic Ocean. *Ciklon-3* (*Cyclone-3*) lost contact with mission control, which was trying to establish whether it had crashed on Wrangel Island northeast of Siberia. Not only was this a blow to Russian plans to modernize its early warning defense system, but it was another reminder of the decrepitude of its missile inventory. *Ciklon-3* was a low payload military rocket, designed in 1977 and still built in the Ukraine, where worker morale is low and spare parts are regularly stolen.

In a report in the newspaper *Segodnya*, a local activist considered taking the case of harmful space debris to the Council of Europe. He described what happens when the Proton and Soyuz boosters plunge to Earth: "They start burning in the outer atmosphere, then break up into fragments from matchbox size to 15 meters long." A chief local official for the village of Ploskoye added, "Lower down, the unburnt fuel explodes over our heads."[14] The official, Viktor Pakhomov, said that not all inhabitants are given the recommended two days' warning. "Military launches are still top secret, but we still know that exactly nine minutes after blast-off the rocket goes over us. There is a terrible noise and fiery explosions."

Danger can also arise when people deliberately break up fallen space debris in search of valuable hardware. After the explosion of a rocket near Baikonur in 1970, Soviet troops found a nuclear battery in the wreckage[15] and actually used it as a hand warmer in their poorly heated guardhouse. Kazakh locals have been known to salvage rocket remains to build primitive houses, but the metal remains are contaminated with highly toxic heptyl. Titanium alloys are used for shovels, roofing patches, and toboggans. In the Brazilian city of Goiania in 1987, four people died of radiation poisoning after busting open an abandoned radiation therapy unit containing 100 grams of the highly radioactive caesium-137.[16]

A particularly troubling scenario is the plutonium from crashed spacecraft getting into the hands of potential terrorists. Although small in quantity, the plutonium in most cases could be used in conventional weapons. Otto Raabe, of the Institute of Toxicology and Environmental Health at the University of California in Davis, admits that plutonium is a potentially hazardous material, even though the levels emanating from the small amounts deposited in debris crashes might not be dangerous.

THE CASSINI CONTROVERSY

The greater menace, Raabe says, is from plutonium dust. Thus, the anxiety of the protestors against the *Cassini* mission have some validity. The first of the U.S. nuclear-powered satellites were sent up in the 1960s starting with *SNAP-10A*, which was sent up on a test mission on 3 April 1965. Further, the same *SNAP* system (standing for System for Nuclear Auxiliary Power) has been used, for example, on *Thor, Apollo, Pioneer, Viking, Voyager,* and the *Ulysses* spacecrafts, and a doubly powerful one was used in the Nimbus polar surveillance program, so it is hardly novel. However there was a sharp decline in nuclear-powered spacecraft until the Star Wars (SDI) initiative of the early 1980s rekindled work in this field.

Cassini was launched in 1997 on an 180-foot Titan-4 rocket that has a failure rate of 1 in 210. It was the most expensive unmanned vehicle ever launched, about the size and weight of a 30-passenger bus. It carried 72 pounds of deadly plutonium to power its numerous instruments. Specifically, the spacecraft contains three radioisotope thermoelectric generators (RTGs), which produce electricity from the heat emitted by the radioactive decay of plutonium-2138 dioxide, which is about 280 times more radioactive than plutonium-238, the material in atomic bomb fallout.[17] The plutonium rests directly above the fuel tanks of the second stage used to propel the craft into orbit,[18] the most dangerous phase of the mission. It swung by Earth on a "gravity-assist" at 43,000 mph in August 1999.[19]

But defense scientists in Washington were worried. The Americans remembered what happened to their own Thor-Able rocket, launched in April 1964 from Vandenberg, when the simple navigation satellite malfunctioned with its SNAP-9A nuclear energy supply pack on board. It blew up over the Indian Ocean and released about one kilogram of plutonium-238 into the atmosphere.[20]

However, nothing untoward happened in regard to the *Cassini* launch, much to the relief of NASA, which insisted that even a launch failure would be very unlikely to release any plutonium because of the massive protective shielding installed. Even a catastrophic launch accident would not release the toxic fuel either, according to the U.S. Department of Energy, which builds the RTGs. The thick encasement of the material means that it will not vaporize by atmospheric friction but break into lumps large enough not to cause harm to humans.

In fact, NASA put the odds of an accidental reentry of *Cassini* at more than a million to one.[21] In any event, a vaporization of the probe in Earth's outer atmosphere would inevitably dilute any spilt plutonium particles to harmless quantities.

Over the next ten years, NASA is planning three more inner space missions that are expected to be nuclear-powered: Europer Orbiter, Pluto-Kuiper Express, and Solar Probe. All will not be able to use solar power alone because of the vast distances from the sun the craft travels. Neither, being lighter than *Cassini* and thus needing no "gravity-assist," will they need to fly past the Earth.

Even so, the potential lethal disaster of renegade rocket stages or complete satellites careening out of control over populous areas is now recognized to be too regular an occurrence. Six fireballs were seen streaking across Finnish skies on the evening of 18 February 1997, said to be debris from a Pegasus rocket, catalogue number 24065U.[22] In October 1999 a Proton's second stage engine failed, destroying its comsat in the process. And again in February 2000, a Russian booster vehicle, said to be the most advanced of its kind, plunged into the Kazakh steppes and was not found. A piece of Proton more than two meters across crashed into a pensioner's allotment in the Kazakh steppes. This fragment formed part of the rocket that powered the Freget module, which carried a dummy of an interplanetary probe that was successfully launched. But the craft was furnished with an experimental inflatable reentry shield that also may have failed.[23] On 21 November 2000 an American observation satellite named *Quick Bird*, launched from Plesetsk, south of Archangel in Northeastern Russia, disappeared after failing to go into orbit. The Russians believe it finished up somewhere in the Brazilian jungle.

Illuminated objects seen in the skies can even baffle the experts. For example, astronomers at the Kharkov University Laboratory on 15 May 1994 claim they saw a "white object" some 35 feet in diameter. They said it was a "bolide sporting a tail of 180 miles in length," and that it flew north to south over the Kursk Belgorod and Kharkov regions[24] at a speed of 20 mps. It crashed some 23 miles southwest of Kharkov. This was plainly bolide, rather than debris, speed, yet its physical description seemed to resemble a man-made projectile.

However, tests done at the site of the impact revealed curious but inconclusive results. The energy released was calculated to be the equivalent of 760 kilogram of trotyl high explosive. It was only when the scientists discovered among other metallic substances in the debris a "crumpled and exploded tube" some several feet across, that they concluded it was probably a crashed satellite. The technical composition of the metals were largely iron (Fe), and minute traces of copper (Cu), nickel (Ni), and titanium (Ti) could be detected. Yet Dr. Vladimir A. Zakhozhaj, director of the observatory, said the metal lumps were "more appropriate to a heavy tank than a lighter artificial satellite."[25] Possibly, then, it was a fuel tank or some other flange-like appendage or a spent rocket, since it could hardly have fallen from a plane without that plane being in serious difficulties.

In Britain the Home Office, jointly with the Ministry of Defence (MoD), treats the space debris issue as one of civilian defense. They have issued a classified circular dating from 1979 that discussed "the recovery of fallen space vehicles," especially nuclear-powered ones, and spelled out how to deal with the aftermath.[26] The ministries warned of their accidentally reentering Earth's atmosphere. The document stated, "It is for the government to decide whether, and if so by what means, a public warning of danger from radioactivity should be given. In reaching the decision, the need to prevent unnecessary alarm would be carefully considered."

The document itemized the various government departments that would necessarily be involved with crashed space vehicles, such as the Atomic Weapons Research Establishment (AWRE), the National Radiological Protection Board (NRPB), the MoD and senior police and fire officers, as well as representatives from the "NAIR scheme"—the National Arrangements for Incidents Involving Radioactivity. The document also has a section dealing with "non-nuclear debris from space," and advises chief police and fire officers to notify the MoD in the event of space debris landing. The U.S. military has also developed contingency plans for nuclear accidents such as mass evacuations, mobilizations, helicopter searches, and ground sensors.

The term "crashed space vehicles" is an emotive term. It was, ultimately, the ballistic missile aspect of the space debris menace that was so worrying. In the next chapter, we will be looking at the way various agencies, scientific, governmental, and military, endeavor to analyze the large items of metallic debris that so often seem to fall to Earth, to determine whether they do indeed have a military aspect.

In conclusion, a few words ought to be said about the international legal aspect. It appears that the Outer Space Treaty has only one clause dealing with space debris.[27] Article VI merely declares that states are internationally responsible for their space activities and those other non-governmental concerns involved in space projects. There is also the question of liability for negligence in regard to "space objects" causing damage when they land on the soil of foreign sovereign states. Once they become debris, however, is any state responsible for them? Can wrongful intent be proved? Under Article IV (1), dealing with a "Registration Convention" requiring fully functioning "space objects" to be listed, no formal requirement to notify changes to the status of these objects was mentioned, as would be the case with them breaking up in space. Another clause, Article IV (3), says that the United Nations should be notified of objects "no longer in orbit" but only "as soon as practicable." Space lawyer Armel Kerrest of the French Institute of International Space Rights says that any state's satellites that become a victim of a collision must prove a fault.[28]

One irony concerns the fear of NPSs on board spacecraft. If debris causes an accident to a flying NPS and this craft then crashes to Earth, both states—the one that caused the collision and the one whose NPS contaminates the ground—may be jointly liable for the burden of compensation to any victims.[29]

Then there is the general problem of allocating blame for damage to a state's space assets from "space debris" itself. Kerrest says that it is virtually impossible to make "outer space pollution" illegal, since all space objects sooner or later are likely to become space debris.

Finally, we must bear in mind that ballistic missiles are merely targeted rockets. Rockets can themselves be weapons, or they can loft military surveillance satellites into space. If large chunks of them fall through sovereign air space and onto sovereign territory, the authorities with jurisdiction over that territory want to know about it. And they do not necessarily tell the world what they have found.

CHAPTER 5

The Moondust Project

In 1967 a secret defense intelligence operation known as "Operation Moondust" was mounted jointly by the United States Air Force (USAF) and the CIA. Their task was to retrieve fallen space debris—what is now known as "space junk"—from infalling satellites or parts of satellites, and other metallic debris that clearly had an aeronautical and possible foreign origin. The collected documents of Operation Moondust have now been made available to the public. NASA also instituted regular procedures for reports of sightings of space debris "as a result of safety destruction action, failure in flight, or re-entry into the Earth's atmosphere."[1]

Clearly, in the decade immediately after the Soviets' historic 1957 *Sputnik* venture, the satellite industry was in its infancy. Hence, the searches of specialist personnel of the USAF were focused mainly on the spent stages of early ballistic missiles and rocket stages—both American and those of foreign powers. The American defense agencies soon generated large files detailing anomalous sightings of objects falling from the skies, and endeavored via internal analysis to determine whether these could be distinguished from ordinary cosmic objects, such as meteorites.

The contents of these files included intelligence gleaned from the monitoring of foreign press reports. In many instances, this entailed a certain amount of educated guesswork as to the nature of the objects being described. Soon the word "satellite" began to appear in the NASA files and then on the Moondust files.

One report appeared to refer to a meteorite that plunged into the sea near Japan in 1960 and was reported by the Japanese Ground Self Defense Force (GSDF).[2] Another example in the NASA files, dating back as early as 17 March 1960, mentioned an item in the Stockholm daily, *Dagens Nyheter*. The journal referred to a glowing object that was possibly a satellite. It was spotted by the crew of Linjeflyg, a Swedish airline. The satellite was said to be of a type known as "Alpha" and was seen from an observatory in Saltsjoebaden and by ground witnesses at Norrtaelp.[3]

The Americans also took note of a SHAPE (postwar allied military headquarters) report in 1962 referring to a metallic object that fell towards the sensitive east-west German border.[4] In another incident a year later, a light commercial aircraft, "Victor Juliet" (C-FLVJ), struck a heavy object while flying over Niagara Falls in Ontario. The object sheered off the left wing and caused the aircraft to crash, with the death of all on board.[5]

The Moondust files also include secret Department of State telegrams sent from foreign embassies to Washington that were often duplicated or forwarded on to U.S. Air Force bases selected to handle sensitive "space debris" phenomena. For example, a telegram from the defense attaché of the U.S. embassy in Mexico City in August 1967 to Washington concerned a sphere two feet in diameter that was found in the southern region of Chiapas. It was identified as a titanium gas stowage sphere of American origin.[6] Earlier telegrams spoke of a "satellite or similar object."[7] A later series of telegrams in October between the USAF Field Activity Group in Fort Belvoir, Virginia and the Mexico City embassy mentioned that "possible space fragments" (from Chiapas) had reentered in March 1965 and were in the hands of officials in the city of Durango.[8]

Another official telegram also reported in August 1967 that a "cube-shaped satellite" weighing approximately three tons had been discovered some 50 miles from Khartoum, Sudan.[9] In February of the following year a "secret" correspondence between Washington and the embassy in Abidjan also concerned "returned space objects." The following month, March 1968, a Department of State Airgram sent to Washington from the embassy in Moscow referred to an article in *Moskovskiy Komsomole* by Villen Lyustiberg, a science editor of the Novosti Press Agency, debunking flying saucers.[10]

A month later the U.S. embassy in Bogota, Colombia, sent a confidential telegram to Washington that referred to a fallen object that had been dissected by the Bogota Institute of Nuclear Studies. It was said to be 32 by 21 inches in size, and a piece was to be forwarded to Washington for analysis, presumably with the consent of the Colombian authorities.[11]

In September 1969, according to a joint U.S. embassy and Defense Department office report headed "Moondust Space Material Collection," reference was made to another Dagens Nyheter article that mentioned a "spherical space object" 38 cm in diameter and made of brass alloy. Its origins were unknown, but it was similar to an object found near the Swedish town of Hede in 1964. It was assumed that the debris was of American origin, and the Defense Department staff were keen to have the items flown back to U.S. military research institutes for analysis. Swedish scientists at the Research Institute of National Defense said that the fallen objects were of U.S. origin. One was said to be a "spherical pressure vessel for gas" probably used to stabilize a satellite or its cameras.[12] There was some debate in the telegram about whether to "pressurize" the Swedes to let U.S. scientists examine the objects, since "if we don't we have to keep in mind that this case may set a precedent for other space objects of greater interest."[13]

On another occasion a secret State Department telegram sent to Washington from the embassy in Pretoria in July 1970 concerned the "space material collection." Apparently South African police had presented U.S. officials with an irregularly shaped piece of metal approximately 34 inches by 20 inches. It was analyzed by South African scientists and others at the South African Defence Institute. The conclusion was that the object was probably "part of a bell-shaped piece consisting of three metal sheets." It was partly corrugated and made of titanium alloy.[14] In a letter from the National Institute for Defense Research of South Africa sent to the S.A. police in Pretoria, the object was declared most likely to be part of a spacecraft rocket mouthpiece, possibly from the *Apollo 13* landing craft, although it could possibly be Russian. It was made of titanium alloy, typically used in space vehicles needed to withstand high temperatures.[15] The embassy included a Polaroid photo of the object supplied by the U.S. military, who were based near the embassy. The caption on the photo studiously avoided the *Apollo* connection, saying that the object was "presumed to be a space object received from the SA Police."[16]

The U.S. embassy in Wellington, New Zealand, received a telegram from Washington, this time with the authorization from the NORAD space defense center at Colorado Springs, on 6 April 1972. It said that if there was any doubt about the origins of the space object, it must be returned to the United States with haste. The telegram requested U.S. examination and photographs of the object, which was in the hands of the Department of Science and Industrial Research (DSIR) in Wellington. This was reluctantly agreed to by the DSIR, and the U.S. Joint Intelligence Bureau examined the object with spectographs. Again it was found to be made of titanium alloy and weighed approximately 30 pounds. Its diameter was 15 inches and it was a quarter of an inch thick. The DSIR assumed it to be of Soviet origin because of the mix of latin and cyrillic letters on the casing, plus the familiar five-pointed Soviet star,[17] and the predictions of reentry of a satellite emanating from the Caucasus region.[18] The Soviets denied this.

The question of sovereignty over debris found on foreign soil was a tricky one, which the Americans endeavored to treat with circumspection. In most cases the U.S. defense establishment replied to the intelligence messages received from their embassies. Its concern was to capitalize on its own advanced knowledge of both space science and of ballistics. It believed there was a serious defense issue at stake, and would be ruthless in insisting that space debris be examined by military experts based in U.S. embassies, or that the objects be sent by military aircraft to America for analysis.

The Moondust files were largely phased out in the late 1970s as other monitoring agencies took over, in particular the Center for Orbital and Re-Entry Debris Studies, part of the Aerospace Corporation in Los Angeles. At the center, mass spectrometer analysis is done to determine the origin of the junk, since different countries use different alloys, and such analysis can also help American engineers better understand how metallic materials perform in space. Intact man-made objects continued to fall to Earth during the 1980s, with a slight decline during the early 1990s.

But several troubling events happened in the year 1997. A cylindrical jettisoned fuel tank from a Delta-2 rocket weighing some 260 kilograms and measuring three meters in length, made landfall in January and hit Georgetown, Texas. Its broken remains were taken to the Los Angeles Center to join a collection of other "moondust" objects. Another object, a beach-ball-sized sphere of titanium, which once contained pressurized gas used to force fuel into a Delta-2 rocket engine, also landed in Georgetown, and probably other spheres from the same rocket lay undiscovered in the Texas country-side. In the same year other spacecraft debris was found in neigh-boring Oklahoma.[19]

CRASH-LANDING METAL SPHERES

The Moondust story comes right up to date in the year 2000, when considerable media publicity was given to the strangely intact metal ball, similar to the one found in Texas in 1997, but larger, that was found on the ground by two vineyard workers in the western Cape Province of South Africa. The 33-kilogram titanium sphere was seen to crash to Earth near a town named Worcester in late April. Pieter Viljoen, who owns the vineyard, said his men heard two loud bangs, almost certainly sonic booms, and turned to see the "white hot" ball descending at a tremendous speed. It buried itself more than a foot into the soil. On the same day, a large cylindrical stainless steel tank about the size of a 20-liter paint drum and weighing approximately 260 kilograms with melting rubber on one side, landed on a dairy farm near Durbanville. Yet a third object, a conical section of a combustion-exhaust nozzle made of composite materials, fell on the same day near Robertson and was 60 centimeters long and 30 cen-timeters in diameter at the base.[20]

Nicholas Johnson, chief scientist at NASA's orbital debris program, whose name will frequently crop up in the next chapter, told a Jo-hannesburg radio station from his office in Houston, Texas, that it was likely—in view of the timing of the landings and the size of the objects—that titanium debris had been discovered from a Pegasus satellite or its Delta rocket, launched in 1996. (It was, in fact, part of a Delta-2 launcher used to put a General Positioning Satellite [GPS] in orbit on 28 March 1996.) He said the debris was expected to land in the southern African region, thus confirming that ground con-trollers cannot always steer spacecraft debris into the oceans or suc-cessfully deorbit them, and that debris does not always disintegrate in the lower atmosphere. "Although no one had been hurt by these objects in the last 40 years, the US government would compensate anyone who is struck by the objects," he said.

He also pointed out that the time that had elapsed between the first impact in Durbanville and that in Robertson could possibly have been as little as four minutes, hinting that the debris was falling at some 1,400 kph in its terminal phase.[21] A similar event occurred in 1979 in the Cape Province when two metal spheres were found on a farm in the Kouga mountains. Eyewitnesses reported that they saw three bright streaks of light followed by a loud bang and "an unearthly rumble." These could also have arisen from a deorbiting Delta launcher, according to Willem Botha in his report to the 2001 Space Debris Conference.[22]

THE UFO INTERPRETATION

Many instances exist of missiles and man-made flying objects, plus various items of metallic debris, that have entered Earth's airspace and dangerously crossed the flight paths of aircraft, both military and civilian, as we have seen. What is worrying is that clearly a great many "sightings" of UFOs are man-made projectiles, drones, and ballistic missiles that seem to have gone astray or have illegally entered sovereign air space.

One must always assume, in the absence of proven scientific evidence to the contrary, that lights seen in the sky have mundane explanations. As we have seen in Chapter 3, trails of light have been identified as simple space junk burning up on reentry. Many other sightings are simply spent rocket stages that have been deorbited back to the surface under mission control guidance. To take just one example: one observer of a reentering object from the space shuttle in 1984 reported that with binoculars he could see it flashing and glinting as it fell into the ocean over Hawaii.[23]

However the "lights in the sky" syndrome, which has been with us now for many decades, ought to receive more attention from space and defense scientists than they do. Yet clearly the occult-like nature of the subject has encouraged scientists to give the subject a wide berth.

The Russians are an honorable exception to this "no-go" approach to "UFOlogy." As a result of witnesses reporting "an enormous star, sending pulsed shafts of light to the Earth" near Petrozavodsk, a city in northwest Russia, in September 1977, the Russian Academy of Sciences and the military decided to look into the matter. They wanted particularly (and rather naturally) to know whether there was any link between UFOs and military exercises, or even of the penetration of Soviet airspace by foreign missiles.

The program employed the massive resources of the army plus

most of the scientific establishment concerned with the natural and geophysical sciences. Since the 1960s the Soviets had taken the UFO phenomenon seriously since observations, as in the United States, had also been made by meterological stations, the Air Force, and even by Academy professors themselves. The subject simply could not be hushed up or brushed aside. The general conclusion of a report published later was that both natural phenomena and space debris had been seen, including rocket launches and various kinds of aerospace tests. The causes of illuminated aerial phenomena included the dispersal of sunlight in clouds of dust and gas formed from fuel combustion products. The effects were most noticeable at twilight, with the sun glinting on metallic objects. The report concluded that such phenomena could be seen 100 kilometers above the Earth, so were seen across wide areas.

The sighting at Petrozavodsk was due, the report concluded, to the launch of the satellite *Kosmos-955* from the Plesetsk spaceport. Similarly, a UFO that looked like a giant dolphin, seen from European Russia on the night of 14 June 1980, was caused by the launch of *Kosmos-1188*. The one seen on 15 May 1982 was a *Meteor-2*, and on August 28 of the same year, a *Molniya-1*. Similar effects were seen from the Canary Islands at the time of ballistic rocket launches by U.S. submarines.[24]

In the west, however, there is still much mischief-making and obfuscation. Timothy Good, a leading exponent of the "UFO coverup" school, has concentrated on near-miss events that have either been mentioned in the press or culled from official files. For example, we learn that in June 1985 a Chinese Civil Aviation Authority (CAA) Boeing 747 on the Peking to Paris route "nearly collided" with an object that flew across its path at an altitude of 33,000 feet. Two months later, another near miss occurred on the Italian-Swiss border involving an Olympic Airways flight from Zurich to Athens.[25] A freight aircraft collided with an object on 24 August 1984, and an investigation was launched by the CAA. It was found that the left propeller had struck an unidentified object, thrusting it through the roof and window. There was considerable damage to the fuselage and engine, ailerons, and rudder trim cables. It was reported that "three pieces of a foreign metallic object were found, including a small cylindrical magnet."[26]

Literally hundreds of incidents like this, possibly involving space debris or renegade drones, are cited in the UFO literature, most being followed up with CAA internal enquiries, like the near miss reported by the pilot of an Alitalia plane as it flew over the Kent coast on 21 April 1991. Another eyewitness said the object narrowly missed a passenger jet as it made its descent toward Newcastle airport. The Newcastle *Echo* suggested that the object was part of a disintegrating *Pegasus* satellite.

Jenny Randles and others, such as Nicholas Redfern, also cite press and official sources and talk constantly of "cylindrical slender gray missiles," or "dark wedge-shaped objects."[27] Randles said that on one occasion, aircraft radar had detected an unidentified missile or object that crossed the path of two planes.[28] In another incident a "dark cylinder" was observed by the crew of an Olympic Airlines Boeing 737 on 15 August 1985,[29] while a "cruise" type missile about 15 feet long nearly struck a Cessna plane flying near Sonherham in Sweden.[30] A "gray wingless projectile" was said to have zoomed past Dan Air flight 4700 from Gatwick to Hamburg on 17 June 1991.[31]

Indeed, many journalists suspect that much aerial phenomena has naturalistic explanations. For example a recent issue of *Alien Encounters* said that a reconnaissance satellite, *MSTI-3*, mysteriously went out of commission during a period when the "Phoenix Lights" in Arizona were seen in the month of March 1997.

Still, many other "UFOlogists" have directly challenged the sincerity and competence of air defense organizations and the security services. However, they display an unfortunate ignorance of the technical difficulties involved in antimissile defense systems. Not only must distinctions be made between normal aircraft phenomena and space junk, but also between meteorites, ballistic decoys, and even real enemy warheads, the latter two being dependent on Global Positioning System (GPS)-guided command and control networks. Further, the function of, and the responsibility for, distinguishing between civilian, military, and astronomical aerial objects must necessarily be demarcated between different authorities. We will see in Chapter 7 that satellites thought to be in orbital decay were monitored from French skies by both the civilian Centre National d'Études Spatiales and a military-based directorate.

Missiles with a military function may be a threat to commercial Low Earth Orbit comsats, and space debris and meteorites themselves a threat to such missiles, invoking the nightmarish image of a "space war" or even a nuclear war being started by mistake. It is worth remembering that throughout the Cold War, space-based and ground-based sensors kept watch for signs of offensive ballistic missiles heading toward North America. This mission continues today in, for example, the Defense Support Program (DSP), and comes under the broader heading of Integrated Tactical Warning and Attack Assessment (ITW&AA). In Chapter 7 we will also read about other space object monitoring regimes, such as USSPACECOM, NORAD, the "Haystack" radar, and the Long Range Tracking and Instrumentation Radar (ALTAIR).

So while airports would be under the civilian supervision of the relevant national Aviation Authority, near-space monitoring, because of the importance of ITW&AA, especially, implies that airport flight controllers would more likely pass on any anomalous aerial information to the military authorities if there were the slightest hint that defense issues were involved.

In Britain the Civil Aviation Authority's Safety Regulation Group handles UFO-type phenomena observed by pilots and eyewitnesses. But it is increasingly being subordinated to the authority of the United Kingdom Air Operations Centre (UKCAOC), responsible for the overall coordination of the Air Defence, Ground Attack, and Maritime Air elements of the Royal Air Force (RAF) together with the air forces and navies of its North Atlantic Treaty Organization (NATO) partners. It is an attempt to overcome the military/astronomical/UFO demarcation mentioned earlier. It is from the UKCAOC, for example, that reactions to UFO-type phenomena are increasingly passed on mainly to RAF authorities, who process information received from both civilian and military round-the-clock radars. There are also digital links with NATO partners.[32]

Further, as other nations acquire space launch, sophisticated guidance, navigation and control (GN&C) capabilities, and even nuclear-tipped missiles, the monitoring of aerial phenomena becomes ever more problematic. On the other hand, U.S. national security policy has tried to ensure free access to and passage in space in accordance with international agreements made after *Sputnik* was launched, when the Soviets set as a de facto precedent the right to overfly other countries.

BRITAIN—THE MILITARY INTELLIGENCE DIMENSION

In Britain the UFOlogists' disparagement of official reactions to anomalous aerial activity is due perhaps to the more prominent role given to military intelligence agencies engaged in UFO investigations. The Air Ministry's Directorate of Scientific Intelligence (DSI), which has extensive files on former Soviet rocket research, is part of the intelligence community (the SIS). The DSI came into being through the amalgamation of the intelligence divisions of the three armed services, although each department retains its own investigative teams. The main body for initiating enquiries into suspicious aerial phenomena is the Joint Intelligence Committee (JIC), which in turn supervises other advisory groups and sets the country's intelligence agenda.

This is why it is wrong to imply that the missile detection regimes, not to mention individual aeronautical or defense scientists, are participating in some sinister conspiracy to prevent the world knowing about "alien spacecraft" that have crashed but have been furtively hidden away. "UFOlogists" imply that there should be no secret files at all in defense bureaucracies and that the fact that such files exist is somehow conspiratorial and against the public interest.

The files of the Ministry of Defence (MoD) and the Air Ministry show conclusively that the overwhelming number of UFOs are of terrestrial origin. Jenny Randles complains that no special working party has been set up by the CAA to investigate strange aerial sightings or to distinguish these from "near misses." As we have seen above, she is not quite up-to-date in this regard. This leaves the field open to UFOlogists' wild speculations, she suggests. She cites the drone observed in April 1984 near the USAF base at Lakenheath in Suffolk that was misinterpreted as a UFO, a belief the MoD apparently encouraged.[33] In another case a drone was seen in October 1968 over the village of Boosbeck, near Cleveland, which was "dark and torpedo-shaped."[34]

The switching around and reclassification of files, instead of being the inevitable result of bureaucracy, is portrayed as suspicious, underhanded, or "covert." Redfern says that files from the Air Ministry at the public record office at Kew came from a secretariat within the ministry known as S6. He says this fact was the result of a decision to "re-write" the "Standard Operating Procedures" concerning UFOs, and to split up reports from civilian and military sources and distribute them between S6 and another department known as A.I. (Tech) 5(b). Some MoD and intelligence sections are accused of being deliberately more secretive than others, or of clandestinely operating out of Whitehall and of transferring files elsewhere under different nomenclatures.

The recovery of crashed objects by special teams of operatives, presumably working for security or defense departments, is also supposed to be evidence of a cover-up. Redfern draws attention to mysterious squads who rush out to collect objects of the type documented in the Moondust papers that have fallen to the surface. But even if the parts of craft found were civilian prototypes produced by obscure teams of inventors, there may be legal reasons involving patents or territorial infringements why the recovery team cannot reveal what it has found. It should also be quite clear why Moondust-type objects, because of their overlapping military and space research characteristics, should be subject to official secrecy.

It is of course possible that UFO writers are purposefully being misled by official deception and misinformation campaigns—of which the UFOlogists themselves claim to be aware—precisely because of the potential security issues involved. In other words it is easier to allow the belief in crashed UFO cover-ups to be propagated, in the certain knowledge that such suggestions will not be taken seriously by non-UFO researchers.

Nick Pope, another UFO writer and formerly a desk officer at the MoD, treats the conspiracy issue more mundanely. He hints that many MoD enquiries are indeed concerned with space junk, pure and simple. There are no secret departments investigating UFOs, and the passing around of files represents the "extensive circulation of information among specialists, so that Ufo data can be examined by all those who might be able to explain a particular sighting." Specialist divisions have a particular knowledge of radar signals and satellite tracks. Other interested parties, says Pope, include people at the Ballistic Missile Early Warning Centre at RAF Fylingdales, which is also concerned with space junk.[35]

There are other organizations coming into existence that try to put the UFO sightings within a scientific research perspective. In America the National Aviation Reporting Center on Anomalous Phenomena (NARCAP) is a nonprofit scientific organization based in Boulder Creek, California, that provides pilots and air traffic controllers with a confidential communications access for reporting sightings to qualified and sympathetic listeners. NARCAP employs teams of open-minded specialists in aerodynamics, flight control systems, meteorology, and astronomy. Its website is www.narcap.org. Richard F. Haines is NARCAP's chief scientist and a former researcher for NASA Ames. NARCAP is not associated with any airline or government agency, he stresses. He explains why the new organization was formed. "We conducted a thorough, 50-year review of pilots' reports of unidentified aerial phenomena in America and found almost 100 incidents that appeared to impact aviation safety whether directly or indirectly."[36] Many reports of near misses by pilots suggest to NAR-CAP that unexplained aerial phenomena (UAP) do exist and deserve serious scientific study.

In the meantime, specialist government and defense agencies, plus the early warning systems, have nowadays been incorporated into an enormous investigative industry. As we shall see in Chapter 7, calculating both the size and number of "objects" in near space, trying to understand what the objects are, and differentiating them from "natural" phenomena, has become both an academic and military growth industry. Aided by powerful radar and computers, together with persistence and ingenuity (plus loads of planetary science), the scope and power of what can be done to get a better understanding of both the "lights in the sky" and the space debris crisis has advanced enormously.

CHAPTER 6

The Junkyard in Space

More than 8,000 spacecraft have been placed into orbit since 1957, with the spacefaring nations launching an average of 120 new spacecraft every year.[1] According to Nicholas L. Johnson, NASA's chief scientist for orbital debris at the Johnson Space Center in Houston, rockets have lifted more than 20,000 tons of material into orbit since the dawn of the space age.[2] Most spacecraft, dead or alive, range in mass from 1 kilogram to hundreds of tons. The NASA Satellite Situation Report, in assessing the weight of objects in space, distinguishes between payload and debris. The *Mir* weighed in at 140 tons.

But other spacecraft, including the *ISS*, the Hubble Telescope, and *Kosmos 2372*, to name just a few, sometimes have a mass of up to 1,590 kilograms. The heaviest inactive objects are second stages of the Zenit-2 rockets, weighing up to nine tons each—and there are about 19 in all in space.[3] Some rocket bodies weigh as much as 1,430 kilograms. The satellites, of course, have not all remained in their orbits; earlier craft are continually being replaced by more recent ones. Virtually none of those sent into space before 1980 are still up there.

And the 120 per year varies considerably over time on an annual basis. In many years, such as during the 1970s and 1980s, nearly 140 per year were launched since many had a military surveillance function. The launch rate inevitably slowed after the end of the Cold War, plummeting to 73 in 1996, the lowest figure since 1963, largely due to the greatly reduced number of launches from Plesetsk and Baikonur. In fact, Russia went from 95 missions in 1987 to just 28 a decade later.[4] The shrinkage of Russian space activity has mostly affected low-altitude flights but has had little overall effect on the satellite population. Even so, Nicholas Johnson produced a graph at the 2001 European Space Debris Conference in Darmstadt showing that there had been a 34 percent increase in payloads and rocket bodies in space between 1990 and 2000, and a 19 percent change in the amount of debris to be found in space.[5]

There was also considerable controversy generated in 2002 when the European Union announced that it intended to launch a series of General Positioning Satellites (called *Galileo*) into space to rival the 30 or so American GPS-based ones. The GPS system was already being satisfactorily used by explorers, motorists, and other civilians to help pinpoint their location and guide them to their destinations. In other words, an additional shadow European flotilla of GPS satellites will be orbiting close to the existing American ones. The Europeans' key reason for doing this was to provide what the existing military-based GPS cannot—guarantee of a civilian-based continuous and uninterrupted radio pinpointing service (to be linked to devices such as cellphones).

In the early post-*Sputnik* days, tracking whole satellites was done by, surprisingly, a young team led by schoolmaster Geoffrey Perry of Kettering Grammar School in England using kitchen-sink methods involving mere telescopes, binoculars, and ham-radio equipment.[6] Generations of physics students thus had a thorough practical introduction to this increasingly important branch of applied science. More than once they found satellites that were previously lost in space, or made important discoveries about spacecraft functions that were concealed by satellite owners and the launching powers.

In 1996 the Jet Propulsion Laboratory (JPL) in Pasadena established a new Internet service that could help spacecraft and probes get their bearings in space by calculating the view from more than 25,000 planets, moons, and other spacecraft. This meant that not only would mission controllers be able to find out exactly when a particular celestial body rises into view, but that astronomers could also calculate astronomical coordinates as viewed from anywhere on Earth. Jon Giorgini of JPL says the system includes an enormous amount of orbital data on space objects, natural and artificial. NASA mission planners as well as amateur astronomers use it, says Giorgini, adding that in the summer of 1999 JPL went further by rewriting the software "partly in response to the number of spacecraft-landing projects under consideration." Planners need to know how to orientate an antenna to best pass on signals to Earth, and about the locations of other bodies.[7]

The most automated database, concerned with intact and fully functioning satellites, is the Wintrak catalog, which tracks the craft in real time. Wintrak's current and future tracking techniques can be portrayed in a rich variety of formats, including a 3D globe and alternative Mercator formats, including moving objects. The Molniya Space Consultancy, based in Hastings, England, also compiles complete lists of worldwide satellite launches.

The Traksat site on the Internet provides details of satellite movements, and NASA's Marshall Space Flight Center in Huntsville, Alabama, has a website that shows a Java-based satellite tracker in action. This "J-Track," as it is called, portrays the real-time position of hundreds of orbiting satellites. The U.S. Satellite catalog is another source. The European Space Agency Register of Objects in Space (ESAROS) catalog (of which more later) also follows thousands of orbiting man-made objects through eleven radar stations around the globe. Its catalog has entries concerning object type, and the nature of the manned and unmanned payloads of the launch vehicles. Launch nation activities are recorded, as are the dates of launches of spacecraft and their expected lifetimes, probable decay rates, dimensions, mass, and mission objectives.[8] In addition, the ESAROS catalog is constantly updated to include other global categories of observation and debris detection, and it records fragments of mission-related objects.

JAMMING THE SKIES

The continual launching of space vehicles has proven that humankind has enormous power to change the character of nearby space. Some say there is a long-term problem with NASA's continual ambition, despite the recent setbacks, to launch one probe a month to explore the solar system from 2015 onwards. If each has to be commanded from the ground, the signalling relays will quickly clog the already overstretched array of communication dishes scattered around the world.[9]

In fact, humanity's prodigious launch activities can now actually be seen with the naked eye, especially at dawn or dusk. Rob Matson, an aerospace engineer with the consulting firm Science Applications International in Los Angeles, says that "for all intents and purposes, satellites are mirrors," and can even be seen behind cloud decks. *Echo-1*, long since deorbited, was seen by millions as a bright star, although it was barely 100 feet in diameter.[10] It was so flimsy a structure that the feeblest of solar pressures could change its direction by hundreds of miles.

The Russian space station *Mir* once had the largest visual magnitude, followed by a group of secret American satellites called *Lacrosse*. The brightest object now is the almost completed International Space Station (*ISS*). The Hubble space telescope also occasionally winks brightly in the night sky. The 66-strong constellation of Iridium LLC satellites, many of which will still exist in space for years (see Chapter 3), with their highly reflective flat plate antennas, regularly produced a flare lasting up to 20 seconds. At a peak intensity, a five-seconds-long flash could appear to be 23 times as bright as Venus. A unique feature of the Iridium comsats was that they could be seen from anywhere on Earth as they continued in their orbit.

These bright lights can only proliferate as cheap commercial spy satellites are more and more often sent aloft. Some newspapers publish details of when the brightest satellites can be seen. The more unusual the trajectory, the more likely the satellite is to be seen by the naked eye. The *Near Earth Asteroid Rendezvous* (NEAR) launched in February 1996 used Earth's gravity in a "sling-shot" maneuver to gain momentum on its way to the asteroid Eros. When it zoomed in close to Earth in January 1998 at a distance of 330 miles, skywatchers across the United States witnessed a bright flash in the night sky as NASA turned its three solar panels to reflect the light of the sun.[11]

The light that is bounced back from the metal casing of spent rocket stages or from the rocket's exhaust also contributes to sightings. Scientists at the Smithsonian Center reckoned that sightings of bright luminous rings in Chile in August 1996 were probably the ignition of a Russian VKS Blok 14 upper-stage rocket launched from Plesetsk. Oblique rays from the setting sun had illuminated the exhaust plume, it was suggested.[12]

Arthur C. Clarke also reminded us that in the 1960s, lights in the night skies were caused not only by rocket firings but by three tiny nuclear devices that were launched to a height of 300 miles over the

South Atlantic and injected a huge number of electrons into the upper atmosphere.[13] They were responsible for spectacular auroral displays some thousands of miles distant.

Indeed, satellites are now becoming a nuisance to astronomers. Those using radio and electromagnetic (EM) waves in their research said they felt besieged by the additional satellites launched every year, which tend to leak signals into the wider EM spectrum. So sensitive are astronomical radios that a mobile phone on the moon would be "one of the four brightest radio sources in the sky," according to International Astronomical Union (IAU) General Secretary Johannes Andersen of Copenhagen University. "Man-made radio signals are much stronger than those used by radio astronomers and quite often some of their signals spill over into our quiet band," said Dr. Jim Cohen of Manchester University's Nuffield Radio Astronomy Observatory at Jodrell Bank.[14]

There has been a growing recognition that the commercial use of space must be balanced against scientific use. Derek McNally, of University College, London, pleads for more safeguard measures. Astronomers, he said, are getting paranoid. "There is nothing in the system that is pro-astronomy. We need to form a comprehensive resolution to feed into the UN system that will let astronomers coexist with all the other beneficiaries that use space."[15]

Perhaps it is because astronomers are still building observatories (and would like to build more) that they feel so strongly that, through concerted action, they could go on the offensive in regard to light and radio pollution. They have said they wanted to clear up, for good, the fog of lights, radio signals, and orbiting junk obscuring their view of the cosmos. They have complained about the absence of global surveys of artificial sky brightness. Images gathered from the USAF Defense Meteorological Satellite Program reveal how serious the sky haziness problem is.

There is much discussion about putting GEO craft into "graveyard orbits" about 400 kilometers above GEO. This has been recommended by space agencies like Nasda (of Japan), NASA, and the International Academy of Astronautics.[16] But, say scientists from the European Space Operations Center (ESOC), only about one-third of all satellites follow the internationally agreed recommendations about after-life disposal of satellites at GEO. They pointed out that 58 satellites had reached their end-of-life phase during the last four years up to the year 2001. According to DISCOS, a major debris-counting system, some 38 percent of these were simply abandoned.[17] Over 350 are drifting mostly above GEO, but the majority cross the GEO zone dangerously twice a day, thus causing a further risk of collisions.

The issue has been studied by Pierantonio Cinzano of the University of Padua, who has produced a new map of light pollution that will soon appear in the monthly notices of the Royal Astronomical Society. Cinzano hopes governments will treat the issue more seriously than they have done.[18]

THE ELECTROMAGNETIC SPECTRUM

On the electromagnetic spectrum chart, all radio and TV signals in the AM, long wave, FM, short wave, and VHF bands are measured in Megahertz (MHz), which ranges from about 0.53 MHz to 230 MHz. At the upper end of the spectrum, digital systems for radio and TV, UHF, remote-control car keys, and various military radios, go up to 953 MHz. But when radio waves merge into microwaves, the Gigahertz (GHz) band comes into effect. Gigahertz are used by radar systems and the General Positioning System (GPS) of direction-finding satellites, and various mobile phone systems such as Orange and One2One, DECT, and digital cordless phone systems, as well as skyphones in commercial aircraft (1.25 GHz to 250 GHz). Satellite TV signals in Europe range from 10 GHz to 18 GHz, with even higher frequencies reserved for the next generation of high-speed wireless Internet connections.

One worry for astronomers is the specific GHz ranges where radio telescopes search for aliens. This spot could be interfered with by the proposed G3 phones, skyphones in commercial aircraft, and the Bluetooth channel-hopping microwave range (22.40 GHz).[19]

Hence, the obvious disbenefit of the proliferation of EM transmissions of various sorts. Although they have negotiated that the lower edge of the GHz spectrum and parts of the MHz spectrum be kept for astronomical listening purposes, astronomers have slowly been hemmed in as chunks of the radio spectrum are encroached. Scientists were particularly worried about the 66 Iridium satellites, which were using wavelengths catastrophically close to the spectrum in which signals from common hydroxyl ions could be detected, as they emit radio signals at a frequency of 1612 MHz.

Ground-based telescopes find these signals difficult to discern behind clouds of dust and gas, but they would be easily detectable by the Iridium satellites (as they are, of course, by space research satellites) operating on a neighboring frequency that is billions of times more powerful than the faint hydroxyl signals.

But the Iridium satellites were not sent up as a scientific research exercise, even though they have magnetometers and other devices that give them more of a sense of space awareness than other comparable comsats. In the meantime, even very faint "leakages" from their assigned wavelengths can drown the hydroxyl signals out. Paul Scott, a scientist at the Mullard Radio Astronomy Observatory at Cambridge, said that although his work had not yet been affected by the satellite transmissions, it had caused much bad feeling within the scientific community. Frankly, it seemed as if commercial interests had overidden the concerns of astronomers.

In fact, the clamor raised by radio astronomers forced Iridium to sign a deal with the European Science Foundation in June 1999 to limit its pollution of the hydroxyl waveband within six years.[20] "This is the first time we've gone to the international level to seek solutions" says Woodruff Sullivan, an astronomer at the University of Washington in Seattle.[21]

At a further meeting in Vienna in July 1999 attended by scientists from 25 countries, it was alleged that astronomy, the oldest of all the sciences, was in danger. The meeting was arranged by the International Astronomical Union (IAU) and the findings of those present at the meeting were delivered to Unispace III. Thanks to the efforts of the IAU, the ball is now firmly in the court of the UN Committee[22] that was later set up by Unispace III, and could lead to binding international agreements. Indeed, astronomers won further protection for a chunk of the radio spectrum at the World Radiocommunication Conference in Istanbul in June 2000. Another 90 GHz was added to the 44 GHz already allocated in the millimeter-wave band. Some unused satellite channels were shifted to leave clear windows through the atmosphere for radio astronomers.

The LEO orbital zone is less than 500 kilometers up, and can often be as low as 350 kilometers. The GEO altitude is often as high as 36,000 kilometers. A semisynchronous Earth orbit and GTO transfer orbits are between 10,000 and 40,000 kilometers in altitude. There is always a much greater chance of a satellite breaking up at LEO because at this height, colliding objects need only be 100 times less massive to cause the same type of damage that would occur to a craft at GEO.

At present there is a high concentration of spacecraft in low Earth orbit (LEO), although the number involved varies according to which part of the Earth's atmosphere they fly through. Most of the LEO comsats spend no more than six years in space before they are de-orbited either voluntarily or involuntarily. Some space vehicles, like the *Big Bird U.S.* spy satellites, have only a very short life span of six months or so, hence they plunge to Earth regularly.

THE MOONDUST SYNDROME RESURFACES

In the meantime, scientists of Japan's NASDA (space development agency) are worried about space junk falling to Earth without "complete combustion," as they put it, implying that some objects simply fall as heavy lumps of metal,[23] a concern implicitly expressed, as we have seen, in the CIA's "Moondust Files." A group of European scientists also points to the uncontrolled nature of rocket stage reentry, much of which doesn't burn up. Parts of *Skylab* and *Kosmos 1686* fall into this category.[24]

The Italian National Research Council (CNR) is monitoring orbital decay of "risky" space objects. Alarmed at the German Space Center's (ESOC) information about the uncontrolled reentry of *Kosmos 398* in December 1995,[25] the Italians focused on three other satellites aloft at the time, all of which they considered to be "uncontrolled," including the Soviet experimental lunar lander *Kosmos 398*.

The *Kosmos* omens have never been good. In 1981 the Soviets' *Kosmos 1275* broke up into hundreds of pieces,[26] 30 of them large enough to be tracked. The upper stage of a Pegasus rocket broke up on 3 June 1996, just two years after launch, creating a massive debris crowd of 700 easily trackable objects in LEO.[27] Observers were worried about damage to the Hubble space telescope, which was in the vicinity.

Indeed, in February 1997, the second Hubble servicing mission was threatened by a piece of Pegasus debris that was projected to come within 1.5 kilometers of the *Discovery* craft.[28] A recent NASA study came to the conclusion that disintegrating satellites tend to break apart at about 80 kilometers up while moving almost horizontally at speeds of up to 7,500 yards per second.[29] At this speed, the surviving fragments, including the separation or break-away of larger, intact component modules and boosters, could easily travel another 124 miles per second after a break-up.

The shuttle usually flies about 250 miles up. At this altitude, the edge of the atmosphere acts like a brake to slow down space junk, which usually reaches Earth's surface within months. Greater problems occur at higher altitudes, potentially more menacing for the Hubble space telescope because the virtual absence of atmosphere means that litter can remain in orbit for hundreds, perhaps thousands, of years.

The hazards of spent rocket stages are some of the greatest worries of the space debris researchers, as are separated structural elements such as adapters, pyro-pushers and nose-fairing covers. Some debris consists of apogee boost motors and even entire satellites like the *TV Sat-1*, which was lost after completing its mission.[30] It is virtually inevitable that Geosynchronous Earth Orbit space missions leave second and third stage rocket launchers in orbit at heirarchical altitudes, since booster stages in the higher intermediate and GEOs can take about ten years to fall back to Earth, but those in LEO, or at low altitudes between 250 and 500 kilometers, fall back within a year and get burned up in the atmosphere. For example, Japan's GEO weather satellite, *Himawai-3*, was launched in 1984, but its two upper orbit stages fell back to Earth in only 1994.[31]

According to Nicholas Johnson of the Orbital Debris Office at NASA in Houston, over 500 missions since 1963 have placed more than 830 spacecraft and upper stages in or near the GEO regimes. This zone has been responsible for "collisions or explosions for up to two dozen GEO satellites."[32] He goes on to say, however, that "only two satellite breakups near GEO have been determined with confidence." The first was the *Ekran 2*, which disintegrated in June 1978 as a result of a nickel-hydrogen battery malfunction, but which was not admitted by the Russians until February 1992. The other was a *Titan 3C* Transtage breakup that occurred on 21 February 1992 and was detected by chance during routine Space Survey Network (SSN) tracking.

Further, China's first major upper stage explosion occurred in October 1990 after the second flight of the *Long March 4* vehicle. Preventative measures were put in place after this, so it was surprising that a major fragmentation of a *Long March 4* third stage, weighing 1,000 kilograms and measuring about 8 meters long, took place on 11 March 2000 after being in orbit for five months.[33] The fragments were larger than that from the explosion in October 1990.

Many of the supply servicing modules that visit the *ISS*, now in its last stages of construction, are predicted to destructively re-enter, including the 7-ton Russian *Progress*, the orbital module and the instrument module of the Russian *Soyuz*, the ESA's Autonomous

Transfer Vehicle (ATV), and Japan's H2 Transfer Vehicle (HTV). In addition, says space scientist Jack Bacon of NASA, "several de-orbit propulsion stages (DPS) of the American Crew Return Vehicle (CRV) will be destroyed after separation from the survivable lifting body." This means that over 500 tons of material will either be destroyed in space to end up as debris, and many hundreds more tons of spent booster and rocket stages will re-enter Earth's atmosphere uncontrolled.[34]

In July 1996 the first recognized accidental collision between two known satellites occurred. The French military spacecraft *Cerise*, costing £11 million and with a classified payload, was launched in July 1995 so that the French defense ministry could eavesdrop on radio transmissions at frequencies between 500 MHz and 20 GHz. Although the satellite was owned by the French military, it was built by the Surrey Satellite Technology Group and operated from a mission control center at the University of Surrey in Guildford. French scientists at the Centre National d'Études Spatiales (CNES) reported an important change in the "moment of inertia." While in orbit, the satellite was struck by a fragment from a European rocket body that had exploded ten years earlier, although *Cerise* was still able to continue its mission after admirable efforts by its controllers.

According to the USAF, the rocket body fragment "increased in size," and a new, uncataloged piece of debris appeared "running with *Cerise*."[35] The rocket's third stage, in fact, exploded after it was cast off from its launch in 1986. The debris was travelling at 50,000 kph at the same height as *Cerise*, more than 700 kilometers above the Earth.[36] The rocket struck the spacecraft's altitude-control boom at a speed of nearly 33,400 mph. The *Cerise* collision fortunately did not generate much debris, although it could have done. The satellite went tumbling head over heels when the boom disintegrated. At first the Surrey group had no idea why the satellite was spinning off course. But with the aid of NASA and the U.K. Space Track Network the group narrowed it down to the rogue rocket stage.

Another serious incident occurred in February 1998 when a Minuteman ICBM had to be destroyed half an hour after it had been launched. This missile was to have landed two reentry vehicles in the Marshall Islands, but spiralled out of control, it was believed, after being hit by an item of space debris.[37] A piece of rocket booster also destroyed the final stage of a ballistic missile a month earlier during routine tests of the U.S. military's missile defense system. The five-meter-long object, which had been jettisoned after deploying dummy warheads, collided with the debris about 390 kilometers above the Pacific.[38]

In the meantime, despite the decline in the use of solid-rocket motors in recent years, the most abundant form of debris, numerically speaking, is solid rocket exhaust or effluent, which is compensated for by the growth in power and size of modern spacecraft. These, in turn, expel more exhaust gases. Hence, solid rocket motor effluents are a major hazard, as are nuclear power supply coolants.

Even so, the spent rocket engine, or booster, being the largest of all the items of "space debris," is still considered to be a major hazard not only to people on Earth below its deorbited flight path but to still-functioning spacecraft, including the *ISS*. As time passed, the obvious dangers of reentering space vehicles, or bits of vehicles, were beginning to be openly acknowledged by space scientists and astronomers. The overcrowding of the upper atmosphere so close to Earth is unprecedented, as most of the satellites are all at the same altitude and pass through the same regions of space. Nicholas Johnson says that during the 1990s, about six satellite breakups were detected each year by the SSN program.[39] He reported that 55 explosions had occurred during the 1990s, mainly from HAP (Hydrazine Auxiliary Propulsion System) rockets, Deltas, Zenits, and Long March stages. Some sixteen of them had unknown causes, and seven were the result of a deliberate destruction of the payload. Some 23 others were propulsion-related, and many of these were due to a fault in the Proton Block D small fourth stage motors.[40] Many spacecraft break up due to changes in their ballistic coefficiency as appendages soften and fold or separate. Torque pressures, convection and friction heating, vibration, and pure dynamic pressures play their destructive roles. In many breakups, each would generate thousands of bits of debris greater than one centimeter in diameter.

According to Richard Crowther of Britain's Defence Research Establishment (DERA), based at Farnborough in Hampshire (which has now been partly privatized and is known as Qinetia), virtually every day at least one object returns to Earth from space, "ranging in size from the order of a portable phone to large space station."[41] The U.S. National Research Council warned, in 1995, against the perils of space junk. They said that when in high orbits it could remain there for millions of years.[42]

Some specialists are beginning to ask whether the satellite population has reached a critical density. According to two Texas-based space consultants who presented their case at the 2001 conference, it undoubtedly has. They point to USAF missile tests at target satellites and Department of Defense (DoD) laboratory hypervelocity tests, which prove that "large regions of LEO will soon be unstable." The region between 700 kilometers and 1,000 kilometers will slowly become congested with fragments, they said.[43] Even Arthur C. Clarke wonders whether the space age might have to end just 50 years after it began because of space junk. "I'm afraid that future astronauts will have to fly through orbiting minefields," he said in 1996.[44]

In determining how space debris arises, we must bear in mind that no booster or fuel module can avoid an accidental blow-up. Spacecraft must be designed to withstand severe launch and operational loads. High-pressure tanks used in propulsion subsystems to store helium gas to pressurize the propellant tanks, or xenon gas as the propellant for electric propulsion units, are particularly vulnerable. According to some accounts, spent rockets that have blown up in orbit account for 25 percent of known space debris.[45] During the 1970s the abandoned upper stages of seven Delta rockets exploded while still in orbit, due to the accidental mixing of propellants after the rocket had been shut down.[46] A battery on board *Mariner*-7 to Mars, launched in March 1969, exploded in space when leaking electrolyte acted like a rocket thruster and made the spacecraft spin out of antenna alignment.[47]

Unintentional pressure-related detonation can occur at any time after launch, even up to 23 years later. One hazard to spacecraft is the way liquid rocket propellant sloshes around in the fuel tank at low gravity. Fuel problems contributed to the *Ariane-5*'s launch problems, as we saw in Chapter 3. Fuel sloshing occurred on the Near Earth Asteroid Rendezvous craft while on its three-year trip to the asteroid Eros, making it tumble out of control for more than 24 hours in 1998, according to astrophysicists at Johns Hopkins University in Maryland.[48]

On other occasions, sloshing fuel has been blamed for rockets stalling in midflight or exploding. The physics is not well understood, but it is assumed that under almost weightless conditions capillary forces make the molecules of fuel coalesce into slithering, swirling spheres that can smash into the walls of the tank and fundamentally change its center of mass, or cause air bubbles to occur in the fuel feeder lines to the engines. As spacecraft get bigger and electronics get smaller, with up to 70 percent of the craft consisting of propellant, the dangers from the "slosh" factor become that much greater. There is also the obvious danger of fuel-related fires, which could release fatal toxic fumes. Further, there is more than one rocket engine, and these alternate between solid fuel and liquid fuel. In addition, there is more than one stage to a rocket. Any one of these assortments of components can fail, and often do, but at inappropriate times.

Debris can also be produced by surface degradation under harsh space conditions, which include thermal stress and atomic oxygen erosion. The upper stage rocket that had propelled the *STEP-11* satellite into space in January 1994 blew up two years later without warning. It not only ejected some 700 fresh pieces of metallic debris into orbit, but the remainder of its hydrazine fuel and a high-pressure tank of helium as well.[49] Cheyenne Mountain debris observers concluded that fragments of the graphite-coated aluminum foil from the fuel tank had turned into dipole antennae, making them unusually bright and easy to track with radar.

On 3 June 1996, the HAPS in the upper stage of the Pegasus launch vehicle exploded two years after launch because of excessive pressurization in the propellant tank. It produced more than 700 trackable objects, according to Peter Wegener and his colleagues at the Technical University of Braunschweig, Germany, in 2001. They said it was the worst debris-related event ever. Fragments came from the graphite epoxy over-wrapped aluminum tank.[50]

In fact, toxic fuel jettisoned from rockets is, as we have seen earlier in this book, a growing threat to people's health in many areas, the *Moscow Times* reported recently, quoting Aleksei Yablokov, head of the Center for Environmental Policy.[51] Most launch companies already drain residual fuel from junked rocket stages to prevent spontaneous explosions.[52] Even so, fuel droplets and derelict rocket bodies still with fuel on board remain a hazard, and pollution from space missions is a major source of damage to the ozone layer.

In 1989 the JPR Goldstone radar in southern California noticed a large cloud of sodium-potassium droplets from a leaking reactor coolant that had been part of a jettisoned nuclear core. Haystack radar also detected about 70,000 similar droplets, all about one centimeter across, in a region where it was suspected that reactor cores had been dumped.[53]

Now scientists at the U.S. Federal Aviation Administration are concerned about what could go wrong when the fourth stage of a Pegasus rocket, with its remaining fuel load, goes into orbit. Orbital Sciences, the company that makes the Pegasus XL rocket, says it may have to reconsider its policy on fuel stages.

It has even been debated that nuclear waste could be dumped into orbit. This was suggested by a committee at the Institute of Electrical and Electronics Engineers (IEEE) based in New York, an idea also toyed with by the Russians. It could be launched from remote areas such as Greenland or Antarctica in case it exploded into the atmosphere. In theory the waste could stay in orbit until it ceased to be radioactive.[54] But Steve Aftergood, of the Federation of American Scientists, a pressure group based in Washington, said it would never happen because of the cost of packing it in strong protective containers, not to mention the public outcry that would ensue. The National Academy of Sciences says the containers that would be required for 50 tons of plutonium alone would weigh more than 1,000 tons and cost more than $10 billion to put into orbit.[55]

Mission-related rubbish is generally reckoned to account for more than 1,000 known objects.[56] The most potentially lethal bit of space garbage, referred to earlier by Captain Steven Ramsey, was the outer glove lost by *Gemini-4* astronaut Edward White (who also lost his helmet visor), while he was performing America's first space walk. The hundreds of spacewalks by other astronauts since have resulted in other tools and items such as spanners and cameras slipping from their grasps. When NASA's Compton Gamma Ray Observatory (CGRO) was ditched in the Pacific earlier in the year 2000, its debris spread over 1,500 kilometers and included British-born astronaut Michael Foale's training shoes and toothbrush and Yuri Gagarin's portrait. Some pieces of the debris were as large as a small automobile.

Some American missions also jettisoned sensor covers and altitude-control devices, and some Russian missions could—based on past experience—leave behind more than 60 distinct items of mechanical debris in various orbits. Pam McGraw at the Space Shuttle Mission Control once complained that the Russians tended to throw their garbage, including human waste, out of the *Mir*, in aluminium cylinders about "the size of a rucksack."[57] The capsules are, however, deliberately fitted with burners that propel them into a graveyard orbit, but there is no indication that they are deliberately controlled on reentry.

The crew of *Gemini-4* may have narrowly missed a disastrous collision with a piece of space junk, according to White and McDivit, who said that once on TV they saw an "object" flying toward them. In fact, more than 200 bits of garbage, most packaged in bags, were dumped overboard from *Mir* during its first decade in space.[58] Perhaps the largest piece of broken-away equipment was the upper half of the ESA's multiple launcher (SPELDA) (Structure Poreuse Externepour Lancements Doubles Ariane), with a diameter of four meters and a mass of 300 kilograms.

TINY PARTICLES—LARGE DISASTERS

The danger from even tiny particles of space debris arises from the fact that even the casings of rockets are made of the thinnest possible metal to reduce the launch weight as much as possible, so that the pressures of launch and of the atmosphere enable the sides to flex. To prevent the entire edifice from imploding, the casing is pressurized with air. Hence, a tiny puncture from a bit of space debris risks blowing up the entire thing.

The space shuttle *Orbiter* has received several knocks since it has been in service since 1981. Impact sites have been seen on crew module windows, payload bay door radiators (made of silver-teflon thermal coatings over a honeycomb panel with aluminum facesheets), and on insulation materials affixed to bay door exteriors. According to scientists at Lockheed Martin Space Operations and at the NASA Johnson Space Center, denting has also been observed on a wing leading edge surface, said to come from paint particles and tiny fragments of titanium metal. One piece of paint left a dent measuring nearly 4 millimeters. In another instance in the 1990s, 43 impact sites were found on the radiators, four of which perforated the outer thermal tape and underlying aluminum facesheet. On 54 missions flown from June 1992 to November 2000, 43 suffered debris impacts.[59] In one event, an impact very nearly put a hole in the external manifold that would have caused a leak of freon coolant and drastically shortened the missions. Three windows had to be replaced in the mid-1990s.

Another danger occurred in January 1998, when the shuttle *Columbia* was put into a tailspin by a mere shower of hail. The hail left small holes in the *Discovery*'s external tank (ET). Left alone, ice can form in the holes when highly cooled propellant is added soon after launch. The orbiter could be endangered should the ice pellets break off in flight.[60]

In February 1996 a NASA experiment with the *Columbia* shuttle trailing a thin 12-mile cable attached to a satellite was mysteriously broken, probably by a particle of metal puncturing the cable's thin insulating layer, causing an arc of electricity. NASA engineers have several times had to replace space shuttle windows pitted by flying fragments.[61] A splash of urine, jettisoned by astronauts long ago, was found on the skin of a spacecraft recovered from orbit. According to a report in December 1997 from the U.S. National Research Council, even nuts, bolts, and lens caps dropped by astronauts were a hazard.[62] In addition, there are many clamp bands, explosive bolts, and decoupling springs left in orbit after the satellites have been released from their rockets. In May 1963 some 400 million needles in 80 clumps were released as part of a U.S. Department of Defense experiment.

DODGING SPACE JUNK

Astronauts who actually witness space junk passing by them say it can be an alarming experience. The French astronaut Jean-Pierre Haignere said, when he saw the traces of an impact on the solar panels of the *ISS* module *Spektr*, which had been punctured: "I said to myself on that day, only a miracle had saved the team living there."[63] The ESA had calculated that, over a ten-year period, there was a 19 percent chance of an inhabited module of the *ISS* (what they call the "debris-critical section") being hit by an object between one and ten centimeters in diameter.[64]

NASA and its international partners were embarrassed by the failure of their emergency procedures after a large piece of space junk—a spent Russian rocket—hurtled toward the *ISS* in June 1999, missing it by no more than seven kilometers. Ground control directions from shuttle mission-control engineers, who spend much of their time monitoring space debris in order to predict collisions and can generally give the crews 36 hours' warning of an impending strike, gave directions to move the *ISS* out of the way. Eventually the debris passed away uneventfully. This was because, as with all debris incidents, the odds are greatly in favor of a miss rather than a hit. Mission controllers themselves believed that the June debris event would pass at most within one kilometer of the space station, and there would be just under a 1 percent chance of a collision. "We decided to take a conservative approach and attempt a maneuver," said NASA spokesman James Hartsfield.

One difficulty concerned *Zarya*, one of the components of the *ISS*, which would have had to fire one of its thrusters longer than its onboard computers would have calculated for. The computers hence refused to obey the command, and instead shut down the station's altitude control system. By the time the flight engineers regained control over the drifting Russian module, there was not enough time left to complete the safety maneuver. In October 1999 the *ISS* was successfully pushed into a new orbit to avoid yet another spent booster—this time a Pegasus rocket—drifting across its path. NASA predicts it will have to move the station twice a year to evade space debris.[65]

Eric Christiansen, chief analyst at the Hypervelocity Impact Test Facility at the Johnson Space Center, says: "There's a residual risk that a particle can penetrate that varies from module to module, but there's about a one to two percent chance per module of penetration within a ten-year period."[66] The main xenon tank of the *ISS* that neutralizes electric charges on the station has a 0.45 percent chance of being punctured by debris, which will cause an explosion of the pressurized gas inside, perhaps destroying part of the station. A larger hole could cause a phenomenon known as catastrophic unzipping, a literal implosion in the vacuum of space.

Christiansen adds that the module where the astronauts will live has a 1.8 percent chance of being hit by debris. This does not sound like much of a risk, but the law of large numbers steps in since the modules proliferate as the space station is built up incrementally. With over 30 modules, the chances of penetration rise to 24 percent, and the longer the space station stays in orbit the further the chance of a direct hit goes up. When, for example, Russia began its involvement with the station, all the *ISS* partners guessed there would only be a 19 percent chance of the station being hit by debris.[67] Over 20 years, the chance of a direct hit to a large edifice like the *ISS* goes up even further to about 42 percent, and perhaps about 8 percent for a smaller craft like a comsat.

To calculate the risks to the *ISS* in the early days, the space shuttle crew once attached to *Mir* an Environmental Effects Payload containing an alloy plate, like a metallic fly paper, which was retrieved by another space shuttle and inspected. It was found to contain 38 microscopic craters made by hypervelocity impacts. Another more sensitive plate made of softer materials enabled the engineers to calculate that the impacts were made by both micrometeoroids and manmade junk.

The space junk largely consisted of tiny paint flecks and powdered solid rocket fuel, plus bits of aluminum of the kind that has damaged the shuttle. Oxide particles, formed and released during the burning of solid rocket motor fuels, have also been traced. Larger items, including solar cell glass shards, fragments of embrittled polymer paint binder, and thermal blanket debris, were discovered recently on the *Solar Maximum* satellite.[68]

In fact, trails of minute paint flakes accompany most aging space vehicles, much of it actually facing ahead of the craft itself as it maintains its altitude while the flake particles tend to fall back to Earth. Engineers inspecting a space shuttle after its return from orbit a few years ago discovered that a speck of paint had penetrated five inches into the shuttle's eight-inch-thick windscreen. As a result, when the shuttle is in orbit it is flown upside down and backwards to avoid debris.

Indeed, it is this type of almost inconsequential debris that gives a hint as to the danger that microscopic grains pose, a danger that is vastly magnified because of the enormous velocities that both the spacecraft and the particles themselves achieve. A fragment half a centimeter across travelling at 17,000 mph could make a hole the size of a fist in the shuttle's crew compartment[69] because the combined speed would become 22,000 mph—40 times that of a bullet fired from a .38 Special. A larger object—for example, a shard of metal—would pack a kinetic wallop equivalent to more than 20 times its mass in TNT.

Even tiny objects of millimeter size or less can be highly destructive when they zoom through space. Solar radiation is supposed to push tiny particles further out into space and thus cause them to drift. But they can bunch together in orbits up to 6,000 kilometers above the surface. In fact, space debris is not like the finely orchestrated rings of rocks, dust, and fragments of ice that orbit the giant planets in well-behaved patterns.

Nicholas Johnson reminds us that Earth's satellites and junk "resemble bees around a beehive, seeming to move randomly in all directions," unlike the collisions between natural orbital material that collide at gentle velocities. According to Don Kessler, another space debris analyst, a population of debris items could form a ring around the Earth, like those around Saturn. John Simpson, of Spadus, the space-dust monitoring organization, similarly hopes that we will soon have a clearer idea of what man's detritus looks like from space.[70]

Others feel that the debris would slowly coalesce and may even form one huge artificial moon in low Earth orbit (LEO). Items of debris, then, are scattered into orbits throughout near-Earth space, ranging from only a few hundred kilometers to more than 400,000 up. By analyzing the pock marks found in solar arrays, scientists believe that the distribution of space junk is at its highest concentration in LEOs, below about 600 kilometers. A U.S. military satellite has traced a cloud of orbiting space junk back to a Chinese rocket that exploded on 11 March 1985. About ten times the normal rate of impacts from particles up to 14 micrometers across were noted over an eight-day period as it passed near the orbit previously occupied by the one-ton final stage of China's *Long March-4* rocket.[71]

According to the Colorado Center for Astrodynamics Research, the likelihood of a piece of debris larger than one centimeter hitting a spacecraft during a ten-year period is somewhere between one in 100 and one in 1,000. This implies there will be one hit for every 500 or so satellites. Further, there can be little disagreement on how disastrous spacecraft contact with space debris can be, as we have seen. Even droplets of coolant can be dangerous, since solidity or liquidity at such hyperspeeds makes little difference to the damage potential.

Experiments at the University of Texas with a "light gas gun," which uses extremely high-pressure hydrogen or helium gas to accelerate small pellets to speeds of several kilometers per second, show that even thick aluminum walls of model satellites are no protection for tiny speeding particles. In most cases the models are completely destroyed, with the aluminum walls still hot after the test and the electronic circuits boards charred and disabled.

Similar hypervelocity tests have been conducted at NASA's White Sands Test Facility in New Mexico with pellets five centimeters wide, with similar results. Thick aluminum plates produce a spray of tiny cascade particles. The satellite walls behind the aluminum shields are actually penetrated, to show pits and cracks. The average impact speed in LEO is about 10 kilometers per second. Tests in European laboratories with the use of specially adapted "light gas guns" can reach velocities of just over this, but Japanese researchers complain that the most their tests can reach is just 7 kilometers per second.[72]

The problem of how much damage space particles can do to spacecraft actually prevents scientists from getting a true picture of the hazards of outer space. What bedevils so much of the debate is the difficulty of trying to work out just how much "debris" is up there. In Chapter 8 we will be examining the input of natural, micro-meteoroid-sized particles zooming in from outer space.

But in the following chapter we will be looking at one aspect of space debris that has a large human dimension attached to it—how can we calculate exactly how much "debris" is in near space? Are there simply too many databases involved, and are they all measuring the same thing?

CHAPTER 7

Space Debris: The Data Problem

The space debris debate is bedevilled with, frankly, an overproduction of data and data-producing agencies. The "space debris crisis," the subtitle of this book, is also a crisis of understanding. This is because growing concerns about space debris have led to a huge and increasingly unwieldy technological subculture that is not dedicated solely to monitoring the issue but succeeds in adding in other related launch factors. Specialists from the world's leading industrialized nations, especially those expert in astronomy, high-tech space science, and Star-Wars-type defense projects are closely involved.

Almost inevitably a number of discrepancies and disagreements about the true nature of the space debris threat have emerged. Too often methodologies and definitions fail to match up. "Databases" listing and defining the myriad number of objects in space are notoriously at odds with each other.

One distinct difficulty arises from what is supposed to be counted—"satellites" or "space debris." When does the former become the latter, and if and when it does are the various registers adjusted accordingly, book-keeping style? One science writer said that NORAD specialists at Cheyenne Mountain, Colorado, do in fact do this—"always at work, deleting the dropouts and logging new births—

launches, explosions and garbage dumps."[1] But does NORAD also coordinate its databases with all the others in this way? In any event, there is much doubt about what is specifically being sent up into space. You can't "delete and log" what you don't know about.

Whether it is because of a bureaucratic scientism that insists on opening up new satellite monitoring files without checking for duplication, or whether the military angle prevents launching powers from being absolutely truthful about what they are sending up into space, the facts are not always clear. One British politician, in the fall of 2001, criticized the United States for "violating the spirit of the 1975 UN Convention on the Registration of Objects Launched into Outer Space."[2] He pointed out that astronomer Jonathan McDowell of the Harvard-Smithsonian Center for Astrophysics in Cambridge, Massachussetts, has noted discrepancies in the U.S. registry, "and it seems that the Pentagon is keeping mum [quiet] as to why." Peter Hain, a British minister with responsibilities for security policy in Europe, said that the actual content of national registers is left to the discretion of the launching powers.

Nevertheless, the discrepancies have recently become worse. Correct orbits are listed for only two of the ten classified (spy) satellites the U.S. launched in 1999 and 2000.[3] McDowell says that three listed orbits failed to match up with the satellites said to be in their orbits. Another four were wrongly identified or were incorrectly listed. The UN Office for Outer Space Affairs confirms this, but says it can do little about it. Unfortunately, the UN registry relies on a treaty that allows long delays in providing data and does not require nations to give final orbits. Some suspect there is a conflict of interests involved. The UN's outer space convention intended that satellite owners could recognize and point out to others, in a spirit of universal cooperation, when space hazards or damage were in the offing. But governments want to know the orbits of other objects mainly for security or defense reasons, according to Charles Vick of the Federation of American Scientists in Washington, D.C.[4] Vick suspects that the Pentagon hopes to evade surveillance from space by concealing the orbits of its spycraft, although, of course, other countries have their tracking data as well, as do private organizations, as we shall see.

IS THERE A CONSENSUS?

In the meantime, the reader, as he finishes this chapter, may not only be confused and worried about the true significance and magnitude of "space debris" but be disillusioned to learn about the way vast research sums are spent to arrive at enormously disparate and highly speculative back-of-the-envelope conclusions. In fact, a kind of Parkinson's law is seen to operate: the more research is done into the issue, the less we seem to know about it. As the following chapter will show, we seem to know more about nature's own natural debris in space than about our own junk.

For one thing, there is a lack of any meaningful consensus about the number of intact satellites, whether fully functioning or not, that remain in orbit, let alone satellite "debris." The expression "man-made objects" is all-embracing, and it is not always obvious whether tiny fragments are supposed to be, or should be, considered part of this debris. A "man-made object" could, quite literally, be either a fleck of paint or a fully functioning space telescope.

Ironically such discrepancies arose from the need for a comprehensive European debris database, which gave rise to DISCOS (Database and Information System Characterising Objects in Space). In 1987 the European Space Agency (ESA) established a Space Debris Working Group (SDWG) that issued a final report within a year, with the database becoming operational in late 1990.[5] President Ronald Reagan's update of the U.S. National Space Policy in 1988 was the first White House declaration to specifically address the space debris issue. A degree exhorting scientists, technicians, and "all space sectors" to home in on the issue and to ensure that design, operations, and experimentation were "consistent with mission requirements and cost effectiveness."[6]

A research program was established by the U.S. Department of Defense (DoD) and the U.S. Air Force Phillips Laboratory to assess how space debris "might pose a hazard to DoD assets." In 1993 NASA contacted its National Research Council, and within just two years an important debris research study document was published. The White House Office of Science and Technology Policy and the National Security Council also set up a special space debris working group. In that same year the first conference took place in Darmstadt, Germany. It included representatives from NASA, the Russian Space Agency (RKA), Japan (NASDA and NAL), the ESA, China (CNSA), and India (ISRO).

In fact, bureaucratic machinations on the European, UN, and U.S. level have grown enormously within just a few short years to deal with this new beyond-Earth's-surface environmental threat (with its additional frightening dimension of "cosmic attack" scenarios). An Interagency Working Group on Orbital Debris was set up as a "coordinating mechanism" on this new ramifying scientific/defense/environmental subject. It continues today to be the principal forum. It is currently led by a growing body of U.S. bureaucracies acting together, including the White House Office of Science and Technology policy, plus representatives from NASA, the Department of Defense, the Department of State, the Department of Transportation, the Department of Commerce, and the Federal Communications Commission (FCC).

Other spacefaring countries agreed to hold joint meetings under the auspices of the ESOC (part of the ESA) in order to harmonize future European debris research activities, especially with European space agencies, such as ASI (Italy), BNSC (Great Britain), CNES (France), and DARA (Germany).[7]

A second conference on space debris was also held in Darmstadt in March 1997, and yet a third was held in March 2001. This latter conference was again organized by the ESOC and was cosponsored by the major national space agencies. The IAU meeting held in Vienna in the summer of 1999 had already had their recommendations concerning the abolition of space junk passed to the Third UN Global Conference on the Exploration and Peaceful Uses of Outer Space (Unispace III), set up in July 1999.[8] In the meantime the IADC is expected to push for the UN to force countries to cut down on space junk by 2004.

However, the published proceedings of the third conference made for a daunting read. The text alone was 900 pages long and comprised some half million words representing the findings of a complex web of 241 individual scientists aggregated into 130 groups of speakers who belonged to something like 50 international and national space science research departments attached either to universities or government defense organizations. The military and defense angle creeps in because of the perceived threat of collisions in space between surveillance satellites and degrading comsats or other debris. Hence, there is a legacy of military observations of space missiles (which to some extent explains the reluctance to comply with the spirit if not the letter of the 1975 UN convention), to which further databases have been added.

One confusion concerns the research scope of the Space Surveillance Network (SSN) and the U.S. Space Command (USSPACECOM), with the former being under the auspices of the latter. But both, together with the North American Air Defense system (NORAD), are part of the DoD or NASA, along with similar, lesser agencies. In Chapter 3 we mentioned the SpaceTrak database, and in Chapter 7 we were introduced to the Wintrak catalog, the Molniya Consultancy, and the Marshall Spaceflight website. There is also the Satellite Situation Report, which NASA makes available online. The U.S. Army's 45.7 meter ALTAIR dish on the Marshall Island's Kwajalein Atoll is yet another debris radar tracking facility. Under construction is still another global missile-detecting institution, based in Vienna, the International Monitoring System (IMS).[9] There may now be as many as 50 tracking institutes worldwide.

As a result, enormous difficulties arise in trying to interpret the findings of this unwieldy and mainly U.S. space debris program. The complicated military/defense/space angle, with overlapping or complementary functions hardly helps, as David Spencer of the USAF Phillips Laboratory outlined in his talk to the Second Space Debris Conference in 1997, when he cited nearly 20 official acronyms.

NATURAL OR MAN-MADE OBJECTS?

These projects, originally run by defense scientists, are continuing with earlier missile tracking technologies—what is known as *collateral*-type sensors such as Ballistic Missile Early Warning Systems using "phased array radars." For example, both the French civilian ESA and the French Directorate for Weapons Systems have monitored "space debris" using these collateral systems. In addition, reentry predictions have been carried out for *Skylab* and *Salyut*-7 space stations since 1979, before the space debris search regimes were in place.[10]

David Spencer of Phillips Laboratory at Kirtland AFB says that space junk would not be a problem for "space assets" (like satellites), if they could avoid or withstand a collision. But it is the enormous speeds involved that are the major problem. The greatest fear is of the "cascade effect," where collisions between super-fast objects create even more junk. The cascade effect, for example, blew up the Russian navigation satellite *Kosmos 1275* in 1981, since one debris scientist said that no energy source on board the craft could have resulted in the 300 pieces of *Kosmos* debris detected. Further, at the height of the Cold War, the United States and the Soviets were conducting tests with "Star Wars" weapons by deliberately destroying decoy satellites with high-speed bullet-like projectiles, to create thousands of fragments.[11]

NASA has said the full effects of "cascading" will not occur for at least 50 years, when comsat growth will have severely congested the

Low Earth Orbits. But Britain's Defence Evaluation and Research Agency (DERA) believes it will be closer than this. Richard Crowther says that his computer models of satellite constellations have altered past predictions and have lowered the timescale. "Rather than the cascade reaction occurring in the next 20 to 50 years, we could expect it perhaps to occur in the next 10 to 20 years, i.e., by 2015 at the latest," he said.[12]

Walter Flury, the leading spokesman for the ESA debris monitoring movement, agrees. He presented a position paper at the 2001 conference alleging that only 6 percent of cataloged objects are operational satellites. One-sixth are derelict rocket bodies, while over one-fifth are non-operational payloads. Payload fragmentation and released hardware amounts to 12 percent of cataloged items, and remnants of over 150 satellites and rocket stages that have been fragmented in orbit account for over 40 percent of the debris population.[13]

Since 1961 a great many satellites have been deliberately dismembered, scattering tens of thousands of fragments large enough to be tracked. Some military satellites were purposely destroyed to prevent them falling into the hands of foreign powers. According to science writer James Davies, the USSR had a reputation for doing this, even after the satellites had completed their missions.[14] The Soviets also developed the hunter-killer series of satellites, which were tested in orbit by maneuvering alongside other satellites to explode and destroy their delicate solar panels and electronics. Debris from the destroyed satellites apparently still remains in orbit.[15]

The Russians had also conducted tests of antisatellite missiles (ASATs) that were designed to home in on their satellite targets and explode, adding to the debris. Some were also destroyed by the Americans as part of military antisatellite ballistics exercises. This happened in September 1985 when a USAF McDonnell Douglas F15 Eagle launched an ASAT. The director of the ESA, F. Garcia-Castaner, said at his welcoming address to the 1997 Space Debris Conference that of more than 8,000 cataloged objects, only about 5 percent are "operational spacecraft," the rest being space debris,[16] implying that only 400 spacecraft are still "live" and orbiting. Estimates of launches since 1957 by a joint Russian, Dutch, and British team recently revealed that more than 3,000 launches had taken place, leading to some 3,600 satellites currently in orbit.[17] Walter Flury says that about 300 spacecraft are operational and orbiting in Geosynchronous Earth Orbit (GEO) zones.[18] However, as most commercial satellites (Comsats) are in Low Earth Orbit (LEO) zones, this implies that there must be, in total, considerably more than 400 satellites presently in orbit in the early 2000s.

An earlier ESA debris count was confirmed in early 2000 by Captain Steven Ramsey of USSPACECOM who pointed out, when discussing the monitoring of the deorbiting Iridium satellites, which the center did in collaboration with Virginia's Network Operations Center, that it was actively tracking—specifically—8,120 objects. "We even track an astronaut's glove from a Gemini mission in the 1960s," he said in April 2000.[19]

Once again, confusion about intact and fully functioning satellites reigns. USSPACECOM's catalog, containing over 24,000 "objects," based on data gleaned from over twelve ground-based sensors using both radar and optics across the world, puts defunct satellites at 20 percent of "space objects," and only about 6 percent as "operational payloads."[20] A separate team of Spanish debris scientists, however, say that since 1961 "more than 100 spacecraft have fragmented into orbit, and more than 25,000 objects have been tracked, of which more than 8,000 are still orbiting."[21]

According to German space scientist Heiner Klinkrad, a leading debris analyst at the ESOC in Darmstadt, a total of 26,500 "orbiting objects" have existed and been tracked by USSPACECOM since 1957. More than 18,000 "orbiting objects" have fallen into the atmosphere by the year 2001. He concluded that the Spacecom catalog, which can see only "objects" larger than 10 to 30 centimeters in diameter, was on the order of 8,500. One term frequently used by SSN/USSPACECOM is similar to that used by Heiner Klinkrad's Swiss-German team—"man-made object," rather than simply "object." But, as hinted at earlier, this can mean intact, fully functioning satellites, those that are "dead," or those that have just broken up. Perhaps the word "object" should be abolished from all space debris discussions.

Nicholas Johnson, the chief space debris analyst at NASA, adds further confusion when he says that of the 75 percent of craft that have completed their missions, about half have remained in orbit or been simply abandoned, left to plummet to Earth at a later date. He says that only 4,500 tons remain in the form of nearly 10,000 large "resident space objects,"[22] monitored by NORAD. And while ESA's DISCOS database details only 8,500 cataloged objects, NORAD's "Resident Space Object Catalog" currently identifies only a few more than 3,000 man-made "objects" in deep space. This compares with

Klinkrad and his colleagues, who maintained in 1997 that "defunct satellites" make up 20 percent of the total, with only 6 percent being operational payloads, although we are not told from which database they are working. Another 16,500 objects are known to have decayed under the influence of "airdrag."[23]

Any science writer, then, endeavoring to draw accurate conclusions about the number of defunct satellites in orbit would have to split the difference between the estimates of two leading debris-watchers—Nicholas Johnson of America and Heiner Klinkrad of Germany who between them talk of "half" and "20 percent"—and come up with a figure of 35 percent. In other words, we must conclude that over one-third of satellites in near space are just uselessly coasting around, without performing any further useful functions for either science or commerce, and presenting a hazard to other operational craft while they do so. Further, our conclusions about the number of man-made items of debris larger than, say, 2 centimeters would present us with a figure midway between SSN's 8,000 to 10,000 items and ESA's 8,000 to 24,000 items. We could thus usefully arrive at an average figure of 15,000 objects. However, a possible maximum figure, as we shall see below, could push this as high as 400,000—a gap so wide as to make it imperative that much further study and analysis be performed if the debris-counting industry is to retain any credibility.

The wide variation in estimates of satellite break-ups and other counts of particle debris exhibited at the 1993, 1997, and 2001 European Space Debris Conferences must be subject to more coordinated harmonization in the future. There must be far fewer speakers at the conferences and fewer space agencies that are allowed to participate. The speakers at the 2001 conference, held in Darmstadt, did show some new emphasis on more computerized use of debris-counting methods and techniques (in other words, trying to analyze what has already been counted, rather than trying to do more counting), with additional welcome attention given to mitigation issues, to natural cosmic particles, and to the legal and international aspects of space debris. Future meetings may hence further diminish the necessity to haggle over debris "numbers."

THE DEBRIS-TRACKING TECHNOLOGIES

The difficulties in getting a precise count of items of space debris can largely be attributed to the vast array of monitoring technologies deployed by an excessive number of agencies. Here again there could be room for more pruning. The Johnson Space Center is at the forefront in space item measurement technologies and debris prediction. Their "Spacetrack" system has been in operation since 1969, with a phased radar base at Eglin AFB, Florida, and most space objects pass within its range. It has a memory of orbital details of most known space objects. There are various other types of debris and intact spacecraft detecting processes involving the use of radar, such as the electronic radar fence technologies. There is yet another catalog called the Satellite Situation Report, which NASA makes available online.

The U.S. navy's NAVSPASUR (Navy Space Surveillance System), which detects and tracks satellites passing through an electronic "fence" from a shaped radar beam, can handle more than 160,000 observations per day.[24] The system is based at Dahlgren, Virginia, and the "fence" is strung across the southern United States, feeding information to Vandenberg. Another phased-array radar is based at RAF Fylingdales on the Yorkshire Moors. It is a huge eight-story-high pyramid with a total power output equivalent to 25,000 100-watt light bulbs. It operates in conjunction with NORAD's PAVE PAWS (Positional and Velocity Extrapolating Phased Array Warning System). According to Squadron Leader Paul McGuire, who recently outlined the role of Fylingdales, "Our primary function is to warn of missile attacks; our secondary function, which takes up most of our time, is being part of the space surveillance network. We can track an object the size of a dinner plate at 3,000 miles. When *Mir* lost a camera which floated out into space, we managed to track it." Fylingdales can track up to 800 objects at any one time, all of which are coded. Some more unusual items include an astronaut's glove, a screwdriver, and six Hasselblad cameras left behind on the moon.

Supplementing the British and American space monitoring projects is the one at the Pine Gap Joint Defense Research Facility in Australia, which, like America's National Security Agency (NSA), is largely a satellite and communications listening relay station. It too in recent years has turned part of its facilities, especially its infrared satellites, into monitoring satellite-originated space debris.

Optical telescopes are complemented with electronic and radar accessories. Radar surveys go up to 2,000 kilometers, while tracking from electrooptical telescopes can penetrate into the GEO sphere.[25] Even so, material in GEO orbits is usually observed by electrooptical means (these have been deployed over several decades) by deep space networks such as the Ground-Based Electro-Optical Deep Space Surveillance (GEODSS) and USSPACECOM. The GEODSS optically tracks objects higher than 3,000 nautical miles, even reaching up to 22,000 miles. The new one-meter Zimmerwald telescope located eight kilometers south of Berne at an altitude of 950 kilometers above sea level has a satellite laser ranging device (SLR) for mainly daytime operations.[26]

Some American space debris telescopes, such as the one based at Cloudcroft, New Mexico, have reflecting liquid mirrors that are extra sensitive to distant light objects. One such telescope noted a satellite that exploded in 1996 and was spotted by the Cloudcroft observatory. Some of the particles were described as being "book sized," and the information incorporated into NASA's flight plan for the space shuttle mission of February 1997 planned to repair the Hubble telescope.[27] The mirrors in turn are coupled with an electronic component called a charge coupled device (CCD), which can amplify pictures onto a TV screen and be recorded on videotape. They can produce up to 200 images per second.

One disadvantage, as with all optical telescopes, is that they also show hundreds of thousands of stars into the bargain. But this is balanced by the advantage that fast-moving flashes of light can hint that an object is tumbling through space, reflecting sunlight as it moves.

Radar, optical, and radio-frequency monitoring—what are known as *contributing* detection processes—are also undertaken. The Space Surveillance Network (SSN), which is part of the U.S. Space Command (USSPACECOM), is in turn part of its Department of Defense. SSN sensors include the contributing type of sensor. The most common is the ARPA series. This is the generically named Advanced Research Projects Agency, which deploys technologies for observing space objects and is included in the ALCOR system—ARPA plus the Lincoln Observatory Radar. Then there is the ARPA Long Range Tracking and Instrumentation Radar (ALTAIR). In addition the Edwards AFB deploys the 18th Space Surveillance Squadron headquarters to search over 17,000 square degrees of space per hour, with overlapping coverage. This technique can filter out stars and detect reflected images the size of a soccer ball.

NORAD—using dedicated and collateral detection techniques—follows the Johnson Space Center in importance, with its Consolidated Space Operations Center (CSOC), itself linked to USSPACECOM. It is based at Colorado Springs and has massive computerized files. The Space Control Center (SCC) at Cheyenne

Mountain AFB monitors space objects with three types of sensors that can observe about 45,000 sightings of orbiting objects each day. A monitoring system at MIT Labs in Lexington, Massachussetts also collects data routinely for objects larger than ten centimeters.

The Americans have, consequently, developed more comprehensive catalogs than other countries of individual space objects and aggregations in both man-made and "natural" space, using a variety of high-tech detection techniques that can identify an object, for instance, no smaller than one centimeter in diameter. The USSPACECOM/NORAD/SSN series of catalogs is the one to which all other debris-watchers make comparative reference in the building up of their own databases. For example, the ESA has two registers: DISCOS and the Register of Objects in Space (ESAROS), both of which largely shadow the USSPACECOM database.

In addition, NASA and the other American debris trackers feed data into computer models, such as ORDEM (for Orbital Debris Engineering Model). The EVOLVE model uses statistics to forecast how the debris will grow over decades to come, taking into account future missions, spacecraft designs (incorporating additional prototype devices), and the way space junk can be kept under control.

The gradual post–Cold War drift away from missile detection to science-based ventures and meteorite-watching is epitomized by the IMS, mentioned earlier. It is already, apparently, yielding a great deal of scientific spin-off, including data about space jetsum and violent storms, shockwaves, and atmospheric radiation. "It's a vast new tool," says Hank Bass, director of the National Center for Physical Acoustics at the University of Mississippi. "For the first time we will have a global system of microwaves listening to the atmosphere of the planet."[28]

A midcourse space experiment (MSC) spacecraft was launched from Canaveral in April 1996 to observe debris. It had a host of sophisticated sensors, including an IR telescope and interferometer, a visible light telescope, a UV telescope, and a spectroscopic imager. Its task is to observe debris in different modes, such as in the visible and IR wavelengths. A large population of centimeter-sized objects was detected at between 700 and 1000 kilometers, which reflection and radar polarization signatures hinted were toxic droplets that could have escaped from the more than 30 nuclear reactors raised into these orbits at the end of the Russian RORSAT missions.

Another space junk counter was launched in January 1999. NASA despatched *Argos*, the advanced research and global observation satellite. It carries a huge number of experiments—scientists called it a "Swiss army knife in space," which includes SPADUS, devised by the University of Chicago (see Chapter 10).

A very tentative, indeed highly inconclusive, general consensus about space debris suggests that the number of man-made objects measuring between one and ten centimeters in diameter in near space is unlikely to be more than about 400,000. This is about all that can be said for sure, since uncertainty, as in the case of satellite objects, rules. A sizeable minority of space experts believe that the upper total is much less than this. Some reduce the number of objects to around 100,000 to 200,000, depending, as suggested earlier, on whether the "natural" debris is included.

But this can still be supplemented with "tens of millions of objects between one mm and one cm," according to Nicholas Johnson,[29] all virtually untrackable and hence unverifiable. The MIT reckons there are some 40,000 objects with a diameter of at least half an inch in near orbit.[30]

ARE THE TRACKERS POWERFUL ENOUGH?

One handicap is the limited "radar energy budgets" (as astronomers often put it) for extremely small fragments in space: simply, not enough power. Ludger Leushacke of the Research Institute for High Frequency Physics (FGAN-FHP) in Wachtberg, Germany, criticizes current debris population density models for producing unreliable data for any object less than 50 centimeters in diameter, in spite of the successes apparently attributed to the DISCOS and SSN detection regimes, and the German device able, as we saw earlier, to detect objects of two millimeters in diameter. Despite such claims, only a few radars are up to the job, he says. Further, he says, "only a few dedicated radars worldwide are capable of detecting objects of centimeter size or smaller in LEO."[31] One handicap is that at GEO altitudes satellites are too far away to be easily tracked, but those in LEO experience variations in atmospheric density that can alter the speed of hurtling debris, turning tracking into a kind of hide and seek.

There is, in addition, the question of how accurate geosynchronous orbits can be determined using radar alone: many GEO signals are buried in noise. Both NASA and the Johnson Space Center in Houston have been using the renowned Haystack radar, operating from Massachussetts, since 1990. Haystack can check into smaller objects, but still cannot detect things like pieces breaking off comsats or determine break-up happening. Haystack data also showed that there is a big jump from 10 centimeter-sized particles at GEO to 1 centimeter populations for orbits between 850 and 10,000 kilometers up.[32] Scientists from Lockheed Martin and NASA Johnson say much of this debris comes from the Russian RORSAT reactors and is believed to be composed of droplets of sodium-potassium coolant.[33]

Indeed Allen Li, an associate director with the U.S. General Accounting Office, told the House of Representatives Committee on Science in June 1998 that the Department of Defense's SSN has great difficulty in tracking particles of less than ten centimeters across, let alone anything smaller. Dan Goldin, in reply, told the committee that the *ISS* was capable of withstanding the impact of pieces of debris up to one and perhaps two centimeters: "Tracking down to one centimeter is a very tough job. It's something we must work on," he said.[34] However, upgrading the surveillance system would require antennae with higher resolutions and improved signal-processing facilities, which would have to come out of NASA's existing budget. Li said that this was impossible.

Some astronomers disagree with Li, since those that specialize in the study of natural space debris such as micrometeorites spend their lives trying to gain this higher degree of resolution necessary to get a true picture of what hovers around in the skies above Earth. They know that space is swarming with high-speed grains of dust that have been shed by larger objects such as comets and meteoroids.

As we shall see in the next chapter, these tiny particles from the outer void are just as much a threat to spacecraft, comsats, and space stations as is man-made debris.

Many would say that the threat is indeed much greater.

CHAPTER 8

The Space Invaders

Satellite engineers are aware that just one particle, one electron, zipping through space at 100 times the speed of a bullet, is all that is needed to destroy one of their finely tuned craft, with its sensitive gossamer-light equipment. A tiny object that streaked across the skies on 10 January 1997 may have come from beyond the solar system.[1] It weighed only a millionth of an ounce and is one of a new class of micrometeorites that are little more than specks of dust.

Just imagine, then, what particles millions of times bigger—say the size of a grain of sand—can do. Tiny particles of dust pose a more serious risk to satellites than huge lumps of space junk, according to scientists from Britain's Oxford Brookes University, speaking at a debris conference in November 2000.[2] These grains are the most common form, in strictly numerical terms, of space detritus that actually reaches our atmosphere. This is the true significance of space debris, the kind that is seen all too frequently as brilliant streaks of light in our night skies.

Mark Littman, an astronomer at the University of Tennessee at Knoxville, said that Earth collects an average of 500 tons of stones, dust, water, and gases from space every day. "Over the four billion years the Earth has been in existence," said Mark Littman in 1998, "we have added 16 million million million tons, but even so we have added less than one percent of the Earth's mass."[3]

Stanley Dermott and his colleagues at the University of Florida have used a supercomputer to simulate the trajectories of thousands of individual dust grains. They found that Earth carves out a small travelling niche, leaving a gap in the dust in front of the planet, with a dust trail in its wake. Dermott also suggested that these "wake" particles might strike Earth at such low velocities that they could settle to the planet's surface without burning up in the atmosphere.

Clearly, most observations of "shooting stars" or even more dramatic fiery events appearing in our skies would have taken place several miles above the ground, but many would have started their pyrotechnics some 600 miles up. If any satellite would have been anywhere near such an "explosion" it would have been destroyed or permanently disabled.

Both the conspiratorially minded and those who suspect the missiles might actually be man-made weapons, not to mention defense establishments themselves, are worried about recent sightings of fiery objects zooming through our planet's atmosphere.

What is troubling is the fact that the reported number of "glowing fireballs" in recent years seems to be increasing. Dr. Duncan Steel of Britain's University of Salford says that at least one potentially dangerous asteroid is being discovered daily, and the rate is increasing rapidly.[4] In just one year many "flaming meteor" or "mystery fireball" or "sonic boom" events are reported in the western media, some of which are included in this chapter and will no doubt be superseded by many more by the time you actually read this book. These occurrences at least signify that those campaigning for a Spaceguard missile defense system (see Chapter 11) to be established are not so easily dismissed as alarmists.

Many small objects have been spotted by geostationary Pentagon satellites 23,000 miles above Earth, using infrared detectors designed for ballistic missiles. Indeed, each year military satellites see dozens of asteroids explode in the upper atmosphere, each with a force greater than the bomb dropped at Hiroshima.[5] Explosions have been observed or detected by U.S. defense satellites around the world between 1975 and 1992, but these were apparently kept secret. There have even been descriptions of "nuclear explosions" over the Pacific in 1994.[6]

In Chapter 5 we spoke of the "Moondust" project involving recaptured fallen debris, and in Chapter 7 we mentioned the growth in the number of scientific specialists who try to differentiate "natural" from "man-made" objects. The Prairie Meteor Network of the Smithsonian Astrophysical Observatory was an early attempt to collect and codify natural meteorite debris.[7] The NASA *Explorer 46*, called Meteoroid Technology Satellite (MTS), also does some useful work.

Even so, there is a likelihood that many "shooting stars" are in fact returning debris from space probes and satellites. NASA's LDEF has on board an Interplanetary Dust Experiment, to monitor the lower size limit of cometary and asteroidal grains, with metal-oxide-silicon (MOS) impact detector panels attached. It shows that the overwhelming majority of particles of submicron to millimeter size at the LEO altitude are man-made. "What is more," said a NASA spokesman at the 2001 Conference, "they are often concentrated in clouds of hundreds of kilometer dimensions."[8]

No one can be sure, in any event, what the debris is. "Metal objects," after all, are what many meteorites actually are. Metallic meteors do not burn up so readily and are prone to give off colored sparks. The reentry of the Russian spy satellite, *Kosmos 2343*, first launched in May 1997, suggested in an official report thought to come from the Home Office, was the type of satellite carrying an explosive charge that allows reentry burn-up of some 200 fragments to be less dramatic and more spread out. But the actual break-up of this satellite was put as 16 September 1997. No definitive explanation was forthcoming even at the end of the year, although the general assumption was that it was an extraordinarily bright meteorite.[9]

SMALL ROCKS FROM SPACE

The recognition of the threat from natural space debris is now taken seriously. Some considered that the "end of the millennium" asteroids, coming in quick succession between 1996 and 2000, were a portent. On 19 May 1996, a large asteroid, named JA1, said to be a "rock the size of Britain's Millennium Dome," came within 280,000 miles of Earth. In March 1998 Dr. Brian G. Marsden, director of the planetary sciences division of the Harvard-Smithsonian Center for Astrophysics, predicted that an asteroid spotted in 1997, code-named XF11, might pass very close to Earth in October 2028, indeed even closer than the moon.[10] XF11 was listed as number 104 on the list of PHAs, or "potentially hazardous asteroids." John-Derral Mulholland and his international space monitoring colleagues pointed out the "strafing" of Earth at Christmas 2000 by a 50-meter-wide rock at nearly lunar distance.[11]

There are thousands more "near Earth objects" (NEOs) on Dr. Marsden's PHA list. Most of them seldom give cause for immediate concern. But it does mean that an object on the PHA list can come within a few million kilometers of Earth and "that the object is large enough to have a global effect." Austen Atkinson even suggests there is "an international conspiracy of silence" to conceal the danger of impacts.[12]

The fledgling International Monitoring System (IMS), mentioned in Chapter 7 as having an Earth-monitoring as well as a missile-detection role, spotted a speeding meteoroid that crashed into the atmosphere over the Pacific on 23 April 2001, where it produced a blast nearly as powerful as the Hiroshima bomb.[13] On 27 October 2001 there was an enormous flash, and explosions were heard along many areas of England's east coast. An astronomer at the London Planetary Forum, Jacquelin Mitton, said these were meteoroid fragments following close on the tail of a comet.[14] An asteroid up to 16,000 feet wide gave the Earth a close shave on 7 January 2002, passing less than twice the distance from the moon to our planet. It was spotted by a NEAT telescope on Mount Palomar in California. Benny Peiser, an asteroid expert at Liverpool John Moores University, said that "such an object could wipe out a medium-sized country if it impacted."[15] In February 2002 NASA's Spaceguard survey said that in the year 2001 over 100 rocks in near space more than one kilometer across were discovered, bringing the definite known total to 587 of the estimated 1,743 out there.

Asteroids and meteors are thought to have been created from the rubble left over as planet Earth was formed. Asteroids are in effect giant boulders; many can be five miles in diameter. It is estimated that more than 2,000 Earth-crossing asteroids (ECA) exist within or just beyond the solar system, larger than half a mile wide in most cases, and thousands more are waiting to be discovered, according to astronomer Robert Roy Britt.[16]

There are a number of smaller meteoroids, up to 5,000 a year, that are often as big as a potato—a few can be tens of feet across—which plummet to Earth and can cause severe damage to whatever they strike, including aircraft and satellites. Some scientists believe that there are potentially millions of meteors out there in near space. Indeed, not a year goes by without thousands of these potato-sized objects actually landing intact on Earth's surface.[17] Press reports frequently give examples of vehicles and buildings being hit by rocks from outer space.

Meteors are themselves thought to be the product of the larger-sized asteroids that have been busted apart. Those meteors landing on Earth's surface (strictly speaking, meteorites) are usually slowed to about 500 mph, but can still cause a giant plume of vaporized stone to shoot up from the impact site, blasting a hole through the atmosphere to eject hot and luminous debris. As it cools, the rocky material drifts back as tiny pellets of stone, which are heated by the air to glow hot pink.

What is important to our discussion is the fact that the speed of any incoming missile, no matter how big or how tiny, is colossal. Many of them, large or small, would hit the atmosphere at 100 times the velocity of a speeding bullet, as we have seen, say about 70 kps,[18] and leave a shock wave that would travel at 20,000 mph. Even the smallest asteroid could release Hiroshima-type energy blasts on their way down through the atmosphere.

A new Richter-type scale claims to measure the severity of asteroid impacts, but is reassuring on only one aspect: the greater the destructive power of an impact, the less frequently it occurs. Called the Torino scale, it ranges from 0 for a near-miss to 10 for a global climate catastrophe.[19] But trying to relate the kinetic energy produced and the destructive power of various sizes and masses of objects is, naturally enough, not a precise science.

The class of missile that could cause widespread damage to a large area of the Earth and even destroy nations (but not the entire Earth or civilization itself) is the half-mile-wide object weighing millions of tons. This ought to be the yardstick by which other missiles can be judged and that is easy to remember: such a missile could arrive every 100,000 years. An object bigger than this can arrive between 100,000 and 1 million years.[20]

According to Lockheed Martin's Sandia weapons and space vehicle research laboratory in California, a half-mile-wide comet, if it were solid and struck Earth, would release energy akin to that generated by 300,000 million tons of TNT. Sandia are now effectively able to calculate and correlate the amount of kinetic energy produced by a space object of any given mass and size, and the amount of damage it can do.[21] They use teraflop computers that can perform 54 million cell calculations using parallel processing. For an up-to-date exami-

nation of the threats of cosmic missiles, one can look up the proceedings of the 1995 Planetary Defense Workshop at www.llnl.gov/
planetary website.

Brian Marsden pointed out that even with current technology, it is
unlikely that the warning time for a "long-period" comet (one that
doesn't return to Earth regularly every few hundred years as the
"short-period" ones do) would be any more than a few weeks. Austen
Atkinson, in his book *Impact Earth*, says that we might indeed have
only a few hours' warning of such comets, or perhaps none at all.[22]

In late 1999 the British government said it was establishing a panel
of asteroid-watchers to advise on the risk of Earth being hit by a
bolide. The astronomer Mark Bailey has campaigned for the establishment of a national center, and told the media, "Asteroids and
comets pose a unique hazard to civilisation—it is unbounded in the
sense that the potential risk is destruction, even extinction."[23]

SHOOTING STARS, SHOWERS, AND STREAMS

We need to explain, at this stage, the complicated physical relationship between comets, meteoroids, and "shooting stars." Some of
the natural dust in the solar system comes from passing comets and
some from the asteroid belt between Mars and Jupiter, where billions
of small rocks are constantly smashing into each other. It is when the
meteorites themselves disintegrate further into much smaller fragments, even becoming dust particles, that they become illuminated as
streaks of light in the night sky.

Many of the micrometeorites themselves interact with the tails of
comets. So "shooting stars" are both a meteorite and a cometary phenomenon. Eyewitnesses often report huge dust trails that follow in
the wake of incoming meteorites as they plunge earthward. The nature of the physical object causing the illuminated dust trails is not
immediately obvious. The action of friction, as a particle descends at
speed through the atmosphere, "excites" the atoms in it.

The connection between comets and meteoroids becomes clearer
when we consider the much discussed "meteor showers" and how
often Earth passes through these. These showers have names and
leitmotifs, and appear seasonally or at certain times of the year. They
are themselves triggered by contact with the debris or gases in the
tail of a large comet.[24]

Take, for example the Taurid stream, which is a vast procession of
boulders hurtling around the sun and out towards Jupiter, where most

of the meteors are. They seldom, when reaching Earth's environs, remain in the "boulder" state. They become so fragmented on their inward journey that they "shower" across the night sky as pinpricks of light, similar to the "Perseid showers" that are said to consist of the debris of the comet Swift-Tuttle.

There is another group of smaller meteors, this time coming from the direction of the constellation Orion, called the Orionids. Because of the enormous speeds at which they crash into Earth's atmosphere—at least 150,000 mph—they often burn up spectacularly. Coastguards were called up after a series of fireballs lit up the east coast of England in late October 2001, leaving a light trail that lasted half an hour.[25]

COMETS—CROSSING EARTH AND BREAKING UP

Comets are thought to have formed from the gravitational effects of the sun as the solar system passed through interstellar dust clouds. In June 1995 the Hubble telescope had confirmed that the icy fringes of the solar system were the most likely places where comets resided, known as the "Kuiper Belt" (a more distant source of comets is the Oort cloud). Objects in the Kuiper Belt are thought to be chaotically orbiting, short-period comets zooming around the sun every 200 years. Many would be much larger than commonly supposed, although the vast majority—the many tens of thousands per year that reach Earth—would have nuclei no bigger than an average boulder. There might be nearly 5,000 of them up to six miles across that have Earth-crossing orbits on a short-period 200-year cycle, which means that every 50 years one of them could be on a very close fly-by indeed.

A comet with an "Earth-crossing" orbit is one that could bring any object into Earth's gravitational field with possible disastrous consequences, since it either misses us or it doesn't. Alternatively it could simply be one that zips in and out of the solar system as it blindly follows its own trajectory. They are occasionally nudged towards Earth, and about 70 of them make return visits to our skies at the end of a long, hyperbolic tether, at intervals of approximately ten years. Some 40 others zoom in from further out and can take up to 1,000 years to return, much diminished in size after their tenuous nucleus is worn away by the sun's rays.

Comets have been seen to break up as they approach the sun, with segments seeming to steer themselves out of the sky. Comets can change direction unexpectedly precisely because they are "dirty snowballs"—in other words, they consist largely of rubble enclosed in ice. When the ice heats up on entry, it could have a burst of gas on either side of the object, which could disorientate it in flight.

However NASA in 1996 traced an object circling the sun that had an elongated orbit matching that of a comet, and even reflected light, hinting that it was made of ice and gases, yet it behaved just like a solid meteorite. This would mean that instead of one giant fireball in the sky, we would see several very small ones spaced out over 24 hours.

Tens of millions of people saw the bright comet Hale-Bopp in the spring of 1997, one of the brightest objects seen by ordinary people without optical aids for decades. To the naked eye Hale-Bopp was twice the size of Venus and was shrouded in a luminous haze; a startling sight. Yet it appeared not to move, only drifting away from its position very slowly over several weeks.

Hale-Bopp's imminent arrival was worrying. No computer calculated, nor could it, its true path. At its peak it was travelling at over 70,000 mph, and its closest approach to Earth was a mere 125 million miles. It was reckoned to be up to 60 miles across. With little warning, it was discovered by amateurs and had given the world only two months' warning of its imminent arrival in our skies.

The astronomer Mark Bailey suggests that we probably see or experience only one comet out of every ten that exist. Sometimes only their radiation can be experienced. For example, an alarming report published in *Nature* in May 2000[26] said that a comet that passed by Earth in 1997 would have been bright enough to be seen with the naked eye, yet it was detected only by Finnish meteorologists using an instrument on the Solar and Helisopheric Observatory satellite. Teemu Makinen, one of the Finnish team, concluded that the failure to observe the comet (1 of 5—the other 4 were detected) from the ground "underlines the the need for full-sky surveillance of comets."[27] Mark Bailey has asked the Particle Physical and Astronomy Research Council for money to set up a robotic telescope in the Canary Islands. It would systematically scan the skies for meteorites, feeding data back to Armagh.[28]

To add to the dangers to satellites, Earth is known to be embodied in the inner edge of a nebulous ring of dust that emanates from the sun—each particle no bigger than a thousandth of an inch across. The ring is 200,000 miles thick and about 300 miles wide from its inner to outer edge. But the dust is diffuse—less than one particle per cubic mile, and hence virtually undetectable. In fact this extra

dust ring was only discovered in the early 1990s via computer sim-
ulations at the University of Florida and the Johns Space Center.[29]
The dust ring is affected by "resonance"—the dust particles become
trapped whenever its orbital period and Earth's coincide, with the
dust nearly catching up with Earth's own orbit around the sun to
finally become attracted by Earth's own gravity. As sunlight tends to
slow these microscopic dust grains, the dust is more likely to fall
toward Earth rather than the sun.

PARTICLES THAT GLOW

Solar radiation, then, as we shall see in Chapter 9, plays a big role
in the shooting star scenario. For example, the gas molecules in a
comet's tail are ionized by sunlight and so carry an electric charge
that makes them follow the lines of the solar magnetic field away
from the sun. But the grains are massive enough to "remember" their
direction and speed around the sun, and can actually drift backwards.
Donald Brownlee of the University of Washington talks of "stardust
in comets," and has had "Brownlee particles" named after him. He
calculates that tiny specks of debris drift down from outer space at
the rate of up to 40,000 tons a year.[30]

It is unlikely that exotic "Brownlee particles" can be distinguished
from any other cosmic dust. Virtually all incoming particles and objects will irradiate in some way. The Oxford Brookes scientists referred to earlier, along with colleagues from the Open University,
also mentioned this problem at the Hypervelocity Impact Symposium
held in Galveston, Texas, in November 2000. They believed that dust
grains vaporize on impact to creat a hot, conducting plasma that can
perturb the electronic systems of spacecraft, "potentially rendering
them helpless."[31]

The micrometeorites come in concentrated bursts rather than randomly in smaller aggregations. But they can yield spectacularly large
displays, leaving huge, luminous trails in their wake as they plunge
earthward, and can be viewed reaching the surface, sometimes at the
rate of one a minute. A swarm of them was seen over San Francisco
on the evening of 9 March 1997, giving a dramatic visual effect
against a backdrop of a shining sun.[32]

Particles can appear as white streaks or even globes that are often
tinged at the edges with other hues, a sure sign of spectrum-mixing.
Both pink and blue are next to each other on the color spectrum.
Blue or bluish-red are commonly seen and appear in episodic fashion.

Dr. Allen Hynek, the noted scientific investigator into postwar UFOs, also said that many sightings were undoubtedly meteors, giving off multitoned colored gases as they burned up—a kind of illuminated static. In December 1994 a meteor was detected by six U.S. spy satellites and was described as a white, "glowing fireball," which was white at the center. Officials at NASA said they had received an unusually high number of "white light" meteor sightings throughout the whole of 1997.[33] And thousands of eyewitnesses saw a "very bright white light" tinged with orange fly over the Sierra Madre mountains in Mexico in December 1998.[34]

Still other witnesses reported seeing a number of falling objects plus explosions, and described white, orange, and yellow trails of smoke. The British Geological Society recorded sonic effects from a region some 100 miles north of Edinburgh. Experts from NASA and the Royal Astronomical Society said there was no stray space debris due for reentry.[35]

Just three months later, a brilliant flash of light was seen over Greenland during its almost perpetual dark season.[36] Reports from trawlermen, corroborated by videotape evidence, said that a flash lasting two seconds was seen. One witness spoke of "a very strong light rolling down from the air. It was like a circle burning, and the air around the circle was very light green."[37] Seismometers recorded a ten-second shock. From descriptions of the size of the steam cloud caused by the impacting object, some 5 billion tons of ice were vaporized. Astronomers from the Tycho Brahe Planetarium in Copenhagen said the object could well have weighed a few million tons. In Bogota, Columbia, four children died when their shanty house was struck by a fragment of asteroid rock.[38]

The meteorite that hit Shandong province in China in February 1997 caused a flash of light that was so bright it turned both the sky and Earth red.[39] A daylight meteor was seen three months later as an elongated fireball and was photographed over the skies of Grand Teton National Park, Wyoming. It reportedly plunged towards the surface at 162,000 mph—the so-called "cosmic velocity"—and reached 4,800°C, glowing white hot. It was accompanied by sonic booms loud enough to rattle windows.[40] On 3 and again on 5 October 1997, both in the late evening, the sky over Liverpool was similarly lit up by a smaller meteorite passing overhead from northeast to southwest.

Scotland also experienced the unusual events. A huge bang accompanied by flashes lit up the sky in the Outer Hebrides. Witnesses on the Isle of Lewis saw clouds of smoke, with falling debris spiralling into the sea. The objects seemed to shower "white lights" on entry. And on the ninth of the same month, a meteor loudly impacted the West Texas desert, and on its descent to the surface was described as being bright as the setting sun. The meteor was tracked by NO-RAD and army helicopters were deployed to retrieve the pieces.[41] Emergency services throughout the southwest and Midlands were inundated with calls after a bright, bluish fireball arched low across the horizon on the evening of 15 March 1998. Just three months later police forces in eight counties received calls from hundreds of witnesses who saw bright blue lights travelling from the region of Devon in a northeasterly direction. Southern and home counties had reports of a bright, colorful object "breaking up." Some witnesses saw a group of smaller lights (presumably the disintegrating fragments), while others saw only one light. A ball of blue light seemed to some to leave a trail (or tail) while others said the trail was "burning." Air Traffic Control had reports of a burning plane coming in to land.[42]

There was a massive asteroid airburst over the Yukon in northern Canada earlier in 2000. The blast, which took place 16 kilometers above the Earth, was equivalent to a 4,000-ton bomb going off and was bright enough to turn street lights off.[43] Late in the year, just after Christmas, a small meteorite caused sonic booms and streaks of light over southeast Australia, sparking off calls to the police.

THE SAND-BLASTING EFFECT

Destructive space phenomena are now beginning to be seen as part of a whole, and the omens are not good. We have the lone zipping electron, the "stardust in comets," and other space dust that can be seen by the naked eye as glowing, multicolored fireballs.

In theory, satellites could be hit by an object 1 centimeter or larger once every 500 years, and this is most likely to take place while in the orbital plane of 30 to 45 degrees left and right of the direction of their travel. According to George M. Levin of NASA and Edward J. Philips, a U.S. major at the Department of Defense, Washington, D.C., space stations could collide with natural space debris about once every 95 years.[44] They concluded that the threat from the debris environment is marginally worse at the 1,000 kilometer altitude.

Dr Aby-Evan, director-general of the Israeli Space Agency, is one of many astronomers who believe that there is a great danger of "satellites colliding with particles no bigger than a grain of sand."[45] Other scientists have found evidence that satellites were being literally "sand-blasted" by space dust particles, enough, according to NASA, "to degrade mission performance or cause mission denial."[46] Tony McDonnell of the University of Kent also describes how meteor storms—of the kind that produce a kind of "stratospheric dusting"—pose a threat to spacecraft.

According to Giles Graham of Britain's Open University, his team of scientists found that tiny microscopic craters found on the solar cells removed from the Hubble telescope showed traces of iron, nickel, and magnesium, aluminum, and titanium, virtually all of which had come from micrometeoroids.[47] Some scientists are more sanguine, and believe that particles with diameters of only a few microns across are unlikely to cause "catastrophic damage," although they can contribute to the "erosion and degradation" of surfaces. But it is not only the size of particles that matter. Nicholas Johnson points out that objects smaller than 0.1 millimeters are a potential threat because their sheer quantity regularly inflicts minor damage on spacecraft.[48]

The high impact force (hypervelocity) of natural micrometeoroid debris—often containing elements of magnesium, iron, nitrogen, and sulphur—is the main problem. Most space vehicles at LEO orbit at speeds of about 7.5 kilometers per second, according to Susumu Toda of the National Aerospace Laboratory in Japan.[49] Not only would hypervelocity missiles, however small, cause "catastrophic damage," they can result in the total breakup of spacecraft.

Despite the fact that this seldom happens, the threat of it looms constantly. In 1995 a platform and other technology used to deploy a satellite received "significant" hits in this way. The shuttle has already had tiles and windows damaged through natural space debris impacts. Indeed, on average, one in eight of the shuttle's windows must be replaced after each mission because of pits gouged by hypervelocity dust.[50] Further, the windows of the *Skylab* Apollo Command Modules were found in 1974 to have been struck by small hypervelocity particles, although this was not realized until 1980. The *Solar Max* satellite, in orbit some four years, had 1,900 holes and dents with diameters of about 35 mm.[51] The Long Duration Exposure Facility (LDEF), launched in 1984 and recovered in 1990, sustained more than 20,000 debris and meteroid impacts while in orbit.[52]

Inevitably, large exposed components such as solar arrays are vulnerable. The new *ISS* is most affected by debris due to its large dimensions (more than 2,000 square meters) and its long mission duration (10 years of operations). It is in a highly polluted orbit, and has a permanent presence of astronauts.[53] One of the hazards with the *ISS* is something known as "unstable crack propagation" in the module wall (which has the effect of "unzipping" the module) and other potential depressurizing events.[54] The ESA mentions studies of hypervelocity impacts, or "catastrophic bursts," on pressurized vessels made of aluminum and titanium.

The biggest problem will be protecting the crew of the proposed *ISS*. At the moment the odds of serious damage are around 1 in 20—far too high.[55] The Hubble space telescope (HST) was launched in April 1990 and experienced about 150 impacts that totally penetrated its solar array, according to Walter Flury, jointly working for the ESA and European Space Operations Center (ESOC) and based at Darmstadt,[56] although he did not make clear over what time span this occurred. One assumes Flury's duration period was between 1 and 2 years, because in February 1997 NASA astronauts, while repairing the HST (mainly a U.S. project in which ESA was a participant), noticed the damage that was caused to the multiple layers of insulation, which was clearly a result of debris impacts.[57]

Some 17 hours of spacewalk videos in 1999 showed that the telescope, which had been orbiting in LEO for a decade at altitudes of 320 kilometers at speeds of 8 kilometers a second, revealed that the spacecraft had been hit almost 800 times by micrometeoroids and space debris. Eight hundred hits over 10 years is 80 hits a year. The HST flies with its telescope aperture behind it to lessen the chance of a mission-critical impact. The majority of the debris hits left impact points less than a millimeter across, but the largest measured a massive 4.7 centimeters. The videos were made during the second shuttle visit to the orbiting telescope in 1997. Comparison with the video taken during the first Hubble visit in 1993 suggests the rate of impacts has increased since then.[58]

On another occasion, two clouds of small particles that could have caused damage to optics and thermal control surfaces were observed from ground telescopes during a period of 100 days in the vicinity of the *Mir* station.[59] When NASA engineers took a look at a solar panel taken from the *Mir* space station, in a scheduled attempt to check into how radiation, micrometeoroids, and other hazards of space would affect the spacecraft, they found that the inside had turned into a glassy brown residue, reducing the panel's efficiency.[60]

One other factor has to be taken into account—genuine "sky static." This is because the two phenomena—static and granular particles—often interact. Some debris-watchers say the effects on spacecraft are of three varieties: (1) Penetration by fast-moving particles,

(2) the formation of plasma charges, and (3) powerful electrical (or current) surges produced from the former.[61] In other words, space debris can turn into—or actually be—another form of electric static. It can create magnetic fields that can disrupt the electronics of space vehicles. If a particle penetrates a solar panel to the other side, the chances are that it will build up a "plasma field"—a cloud of electrical energy that will fuse every circuit in the spacecraft and possibly cause an explosion.

At the subatomic level—at the level of radiation and electromagnetism—every device that has a computer installed is vulnerable. When a meteor particle pelted the ESA *Olympus* satellite in 1993, it destroyed its directional control.[62] A single cosmic ray, which consists of a highly charged atomic nucleus, can cause a stray electric charge to "flip" the binary code within a tiny computer component, leading to the scrambling of a microchip memory. Cosmic radiation from deep space could cripple computer systems on airliners while they are in flight and cause them to crash. Many airlines have now banned mobile phones from being used while in flight.

There is some evidence that a particle-debris strike occurred against *Galaxy IV*, a satellite operated by PanAmSat Corporation, after it experienced a short circuit and rotated out of position. It caused massive disruption to millions of mobile phone users and TV stations. *Galaxy IV* was part of a network of a dozen, and there are many similar networks. If one satellite goes down there could be a resulting domino effect. "The most likely source of damage will not be from a rock blasting a hole in the satellite, but from the creation of a plasma or free electric charge on the spacecraft," said Dr. William Ailor of the Center for Orbiting and Re-entering Debris Studies at the Aerospace Corporation.[63]

THE LEONID SHOWERS

This returns us to the subject of meteors. The Leonid showers, peaking every 33 years, are the fastest meteors of all because the Earth meets them head on. According to Duncan Steel, an expert on solar system physics at Salford University in the U.K., the tiny meteors in the Leonid storm will burn up when they are about 60 miles above the Earth.[64] The Leonid shower can be startlingly brilliant, revealing golden yellow and green-blue colors. The wide publicity given to the Leonid storms in both 1998 and 1999 was partly due to the concern of the satellite operators, who fed back to the world's media.

Both showers were of particular concern to scientists at the Unit for Space Science and Astrophysics at the University of Canterbury because of the highly penetrative nature and frequency of such tiny millimeter-sized particles, which the "storms" produced. NASA observations made from an aircraft flying above the Japanese island of Okinawa during the predicted 1998 peak indicated that there were between 200 and 300 meteors per hour—20 times as many as in a typical Leonid shower.

Throughout 1998 military and commercial satellite operators debated the options for reducing exposure to the Leonid storm. So worried were the satellite owners that they pooled ideas about possible remedies at the Leonid Meteoroid Storm and Satellite Threat Conference in California in April 1998.[65] "This will be the most severe storm we've had in the past 33 years, and there are many more spacecraft up there," said Bill Ailor, director of the Aerospace Corporation. Numerous seminars took place at USAF Space Command and the National Reconnaissance Office.[66] One Pentagon source was prepared to call the odds on the risk of a direct hit, which ranged from one in 100 to one in 100,000.[67]

Italian astronomers at the Instituto ISAO-CNR in Bologna have pointed out at the 2001 conference that the 33-year period Leonid stream produced by the comet Tempel-Tuttle is the most hazardous, since Leonids can reach a velocity of 71 kilometers per second—seven times the momentum of common space debris of the same mass, and they can have nearly 50 times its kinetic energy. They said the Perseid meteoroid hazard also accounted for the failure of the *Olympus* communication satellite at its peak on 12 August 1993.[68]

Atomic oxygen (AO) is present in LEO and can cause erosion and oxidation. According to Neil McBridge of the Open University, in a speech presented at the Hypervelocity Impact Symposium in Galveston in November 2000, the *Olympus* satellite probably did suffer an electrical malfunction at the peak of the 1993 Perseid meteor shower, which eventually led to its loss.[69] The ALTAIR radar is often used to assess the characteristics of the ionized energy given off by the Perseid and Leonid storms, is part of the Kwajalein Missile Range (KMR) mission, and is used frequently by USSPACECOM.[70]

It was suggested that some of the satellites be shut down temporarily. But such is the delicacy of spacecraft electronics that this action carries the risk that operators may not be able to turn them on again. Further, the competition for safe slots means that nudging satellites into safer orbits would not please everyone. Aby-Evan of the Israeli

Space Agency added that the Canadians would be able to alert systems operators to shut down their equipment or change spacecraft orientation so that their smallest face is turned toward the meteor storm.[71]

Some satellites were successfully put into a "safe mode," including two Earth-observers *ERS-1* and *ERS-2*, and the sun-observing *Intelsat* also rotated fragile solar panels so that they faced edgeways during the 1998 Leonid storm. In fact, satellite owners breathed a sigh of relief in November 1998 when orbiting hardware seemed to have emerged unscathed. Barron Beneski, a spokesman for Orbital Sciences in Dulles, Virginia, said that the company's 30 comsats and imaging sats survived "without degradation." The Iridium network also came through unscathed.[72]

Yet valuable knowledge was gleaned from the 1998 storm. When the Hubble telescope was turned away from the expected onslaught, the time was put to good use. Teams from the University of New South Wales in Sydney, NASA's Goddard Space Flight Center in Greenbelt, Maryland, and the University of Alabama in Tuscaloosa won a competition to use the space telescope while it was looking "the other way." It was used to study a quasar and galaxies.[73]

The November 1999 Leonid storm supplied more crucial data. Scientists from two USAF jets from Mildenhall airbase in Suffolk, which flew along the coast of North Africa to the Azores, counted 15,251 meteors over a period of several hours. "That's more than I have seen in 34 years of observing them," said Chris Crawford, one of the science team on board.[74]

The 1999 shower was seen best in the Middle East, and two especially equipped NASA aircraft arrived in Israel on 17 November to join an already huge team of scientists at Tel Aviv University's (TAU) Florence and George S. Wise Observatory. The $2.5 million cost of the operation was sponsored by the NASA Ames Flight Center, the USAF, the Israeli Space Agency (ISA) and Israeli high-tech companies.[75] "This was the first-ever joint airborne-radar observation of meteors," said Dr. Noah Brosch of TAU's School of Physics and Astronomy, who coordinated the ground mission. It was a large-scale, integrated view of a meteor storm using multiple viewing techniques such as infrared UV, optical, and laser cameras, said Dr. Brosch, who detected ten times more meteors than were visible. "The mission reflects the multidisciplinary nature of science today," he said.

"There's a very serious risk to satellites," says David Asher of the Armagh Observatory in Northern Ireland.[76] In the meantime, astronomers from the USAF and Canada also monitored the Leonid shower from camps in the Gobi desert and even from Australia's outback, and relayed the information to USAF website satellite operators. NASA also released a balloon from a site in Alabama in November 1999, which reached a maximum altitude of 20 miles in order to expose a special acrylic material at the peak, designed to attract Leonid space dust. "It works like flypaper," said John Horack, a NASA astrophysicist.

In conclusion, the miniscule nature of space particles and the way they irradiate as they zoom down into the upper reaches of our atmosphere tells space scientists they are not really dealing with anything substantial at all. The Leonid storm of 2001, which yielded the most spectacular showers since 1966, was rich in microparticles. The interrelationship between matter, light, and energy becomes ever more nebulous: matter is, after all, merely confined energy. If we return to the beginning of this chapter we will remind ourselves that a zipping electron travelling at 100 times the speed of a bullet can do as much harm to spacecraft as a grain of space dust. The tiny pinpricks of light that constantly collide with the toughened windscreens of the shuttle are simply varieties of sky static.

In the following chapter, we will be examining this phenomenon in more detail.

CHAPTER 9

The Space Weather Threat

Like weather on Earth, weather in space is extremely variable. Conditions can turn "stormy" in a matter of hours and can last for days. And space storms, like weather storms, can change with the seasons and follow their own cycles. But space weather arises from physical processes that are quite distinct from terrestrial weather. The latter is derived from dense, electrically neutral gas in the Earth's lower atmosphere, whereas the former follows the laws that govern plasma—very sparse gases consisting of atoms with either outer electrons missing or extra ones added. These plasma particles, more importantly, affect Earth's magnetic field.

In August 1996 a conference at Montana State University was warned by scientists that "space storms" could have a devastating impact on society's technological infrastructure. "People are beginning to talk about 'space weather,' " said Clive Dye, who heads the Spacecraft Environment and Protection department at DERA. Gordon Wrenn, also of DERA, is similarly concerned about the safety of satellites.[1] In short, our electronics breakthroughs of the last part of the twentieth century have made us peculiarly vulnerable in ways that could not have been foreseen.

The complaints arising from sky static are similar to those voiced by astronomers, who fear that street lights, radio waves, and micro-waves are blotting out chunks of the spectrum. It is an ineluctable fact that almost every aspect of modern society—from home and of-fice to research centers and defense establishments—is being over-whelmed by a tidal wave of electronic information. The World Health Organization recently referred to "EM pollution," or simply "hot spots." When electromagnetic (EM) fields from various sources mix and match within small areas, for example, in a room or a stretch of road, they can produce magnetic chaos. And because spacecraft and defense issues are also involved, there are literally hundreds of space and meteorological departments attached to universities or de-fense agencies that are studying the problem of sky static. They reg-ularly issue uncirculated reports, especially when military satellites are involved, or publish them in science journals.

Satellites are unusual in that they are not really vehicular systems so much as communication systems. They employ a mix of mechan-ical, guidance, and electronic devices. Generally they are a com-bination of intricate subsystems: propulsion technology, attitude de-termination and control, thermal control, and data communication handlers.

Satellites are highly complex devices that can be maneuvered hun-dreds of miles from Earth, hopefully, and are yet still able to with-stand the extreme conditions of space. They can either be assembled using standardized modules or parts, or can be specially tailored to meet individual requirements. The main body of the satellite is called a bus and consists of several sections. There is usually a payload mod-ule that carries the electronics, so the module varies from type to type. The service module, which is probably the same for most sat-ellites, houses the apogee motor and propulsion tanks.

Satellites are, if anything, becoming more vulnerable. Larger un-manned spacecraft typically have primary and secondary structures made from honeycomb aluminum or carbon fiber reinforced plastic facesheets, bonded onto an aluminum core. Space weather protection for smaller satellites also seems to have been decreased. Miniaturized components have been etched onto coatings and deposited onto silicon wafers, and this means that silicon-based components may form part of the primary structure (these are known as MEMs—microelectro-mechanical systems). Many of these are actually externally-mounted as micropropulsion thrusters.[2]

Difficulties started to arise from the enormous complexity built into satellites and their astonishing delicacy. They are, in short, the most intricate and most expensive pieces of machinery in the world (partly because they are so intricate). Unlike road- or land-based military vehicles, where toughness and a degree of durability are built in, satellites require a totally different type of construction approach. Space engineers feel they are launching an ungainly insect-shaped object covered in tinfoil, consisting largely of thin aluminum, inside which exists a messy storeroom of millions of wires, printed circuits, and computer components.

NASA has long been worried. Most scientists agree that electro-magnetism at the lower levels of energy is generally harmless. Light, radio waves, x-rays, and other forms of radiant energy are transmitted through space and appear on scientific instruments as crests and troughs, just as waves form when a stone is thrown into water. The number of waves per second is known as the frequency, and is measured in Hertz. At the very lowest part of the EM spectrum are radio waves and at the upper end are gamma rays and cosmic rays, operating at a frequency so high (from 3×10 to the power of 20 to 3×10 to the power of 24), and travelling so fast that they can cause damage to living tissue and delicate electronic components.

THE DANGER TO ASTRONAUTS

Nevertheless, when NASA joined in with the Russians in launching the first segment of the *International Space Station* in November 1998, it knew it was taking chances because space is full of Russian roulette–type danger. "Going into space, you can't avoid radiation," said George Withbroe, director of NASA's Sun-Earth program in Washington, D.C. He was part of a committee headed by George Cisco of the Center for Space Physics in Boston. The center reported in 1999 on the danger of solar storms to space station astronauts, who could contract cancer by being subject to powerful doses of radiation.

Indeed, NASA's experts discovered that astronauts who had been aboard the *Mir* space station for several months had been exposed to the equivalent of a few thousand chest x-rays—a pretty powerful dose.[3] In addition, *Mir* seemed, after a check-up in 1991, to show specks of uranium dust on one of its instrument covers. Scientists from the California Polytechnic State University in San Luis Obispo suggested that the uranium could have come from nuclear-powered satellites that have burned up on reentry. Or it could have come from an exploding supernova. Among the possible culprits, as mentioned in an earlier chapter, was a U.S. nuclear bomb test carried out in July

1962 at an altitude of 400 kilometers—the highest known for a nuclear test.[4]

Astronauts working on the *ISS* seem certain to be exposed to even higher doses of cosmic radiation. This is because when a high-energy particle strikes a molecule of DNA, there's a chance it will cause an error in the genetic code that could lead to cancer. The conditions astronauts face are far more severe than for those on Earth. Gautam Badhwar, a scientist with NASA's Johnson Space Center in Houston, is fearful of the effect of cosmic rays and solar particles.[5]

Tracy Yang, also of Johnson, has counted the mutations in the chromosomes of astronauts who have spent several months on board *Mir* and found that they had been subjected to about 0.15 sieverts, the equivalent of a few thousand chest x-rays, while the usual allowance for U.S. civilians is no more than 0.05 sieverts.[6] It is admittedly difficult to determine, given the random way in which cancer afflicts individuals, whether these high doses inevitably mean a higher cancer risk for astronauts.

Gautam Badhwar estimated that a 180-day mission would amount to the risk of cancer being less than one percent.[7] But as the *ISS* is due to stay in space for over ten years, if 100 astronauts stay on board for that length of time, the odds of one of them dying from their radiation exposure are "even."

According to the Institute of Epidemiology in Copenhagen, which published its findings in December 1999, experienced commercial pilots are five times more likely to get myeloid leukemia, with the danger increasing the higher they fly.[8] Above 26,000 feet, the chances of contracting leukemia are increased five-fold. "Every time we fly," said Hans Storm of the institute, "there is some risk, but our study indicates there may also be a risk from cosmic radiation." But the research team stressed that the link was a weak one and applied only to long-serving flight crews.

Nevertheless, the threat is taken seriously by the aviation industry. From the autumn of 2000 and for the following three years, radiation detectors will be on board Virgin Atlantic planes to monitor radiation exposure from deep space. The Earth's magnetosphere and the sun, and its effects on crew and passengers in aircraft, will be studied. According to Bob Bentley of the Mullard Space Science Laboratory in Dorking, Surrey, aircrews are more likely to develop breast, prostate, rectal, and colonic cancers than the rest of the population, although "the nature of the risk to aircraft crew is poorly understood."[9]

Satellites themselves, as hinted earlier—precisely because they are most susceptible to the effects of solar radiation—can act as detectors. The *Iridium* satellites each carry a magnetometer and can monitor space weather. "Essentially," says Brian Anderson of Johns Hopkins University in Baltimore, "we now have a network of weather stations in space."[10] There is also a network of ground-based radar stations called SuperDARN (Dual Auroral Radar Network) that keeps track of the electrical field surrounding the planet.

Even quite small doses of electromagnetic energy these days are considered dangerous. Microwaves, at a frequency slightly higher than that of radio waves (as, for example, when they are emitted by mobile phones), have also been the cause of anxiety about cancers. This fear is generally dismissed, although physicist Gerard Hyland at the University of Warwick says it is possible that microwave frequencies resonate "endogenously" with living beings. Because emissions from human cells are coherent, or in phase with each other, microwaves could disrupt this harmony, and the rate of cell division in living tissue could be affected.[11]

In 1999 the European Union (EU) jointly initiated research with the Aviation Health Institute (AHI) in Oxford to alert large corporations about the radiation dangers of frequent business flying trips. In the same year, Britain's National Physical Laboratory started to collaborate with the Civil Aviation Authority, Virgin Atlantic Airways, and the Mullard Laboratory to measure airflight radiation doses. Aircraft crews are regularly monitored for cancers, and Farrol Kahn, director of the AHI, says regular flyers should be, too. The U.S. Department of Transportation predicted that among 100,000 crew members who flew 960 hours a year, about 30 extra cases of cancer would occur compared with the numbers expected in the population at large.

Travellers on regular long-haul flights could experience the equivalent of more than 100 x-rays a year taking into account latitude and altitude, said Hans-Georg Menzel of the EU's radiation protection research unit. For more than 20 years, John Hall, a music business executive, flew regularly between New York, Paris, Sydney, and London. In 1993 he contracted cancer of the bladder and came to believe that flying was the main culprit. The closer one flew to the poles, the greater the hazard, he said.[12] In comparison, passengers on just one London–Tokyo flight return received the equivalent of six x-rays, and the equivalent for someone on the ground is 13 x-rays a year.

In 2001 there was renewed concern about passengers and crew members flying on the recently inaugurated North Pole route between New York and Hong Kong. High levels of radiation are usually associated with the polar routes, and regular passengers could be receiving doses that exceed the recommended levels set by the EU and the Stockholm-based International Commission on Radiological Protection.[13]

Space crews themselves speak of alarming sensations when inside an orbiting spacecraft. Sergei Avdeyev, a Russian cosmonaut who spent several months on the *Mir* space station in 1992 and 1995, ranked the high solar storms occurring at the time as one of his most terrifying experiences. Weird lights were flashing all around him: "I could feel particles of radiation walking through my eyes, floating through my brain and maybe clashing with some nerves."[14]

SPACECRAFT AT RISK FROM COSMIC RAYS

Many space happenings are potentially lethal to machines and humans alike. Gamma rays were first discovered in the 1960s by American spy satellites in an operation code-named Vela, which was on the lookout for possible Soviet-conducted nuclear explosions in space. By the 1970s, thanks to scientists at the Los Alamos Laboratory, it was realized that they came from a long way beyond our own sun. Indeed, they seemed to come from beyond the flat disc of the Milky Way galaxy itself.

When a gamma ray hits an atom in the upper atmosphere, it dislodges a cascade of electrons that generate tiny flashes of light called Cerenkov radiation. There have been hundreds of theories about gamma rays and their consequences, not only for Earth's biosphere but for the defense and the growing satellite industries. By some estimates they radiate more energy in seconds than the sun does in 10 billion years.

The rays are thought to be the violent aftermath of stars in our own galaxy reaching the end of their lives to explode violently. These newly formed supernovae created a celestial object that was made entirely of superdense neutrons left behind in the neutron star, and it was this that posed a threat to any cosmic object that came anywhere remotely near it. Stan Wooseley of the University of California at Santa Cruz draws our attention to gamma ray bursts (GRBs), which are incredibly rapid explosions of cosmic energy.

Scientists have recently calculated that one gamma ray burst came from a point that was 13 billion light years distant, almost on the edge of the visible universe. The implications were shattering: in just a few minutes GRB 97/1214 had released more energy than a whole galaxy in ten years.[15] Don Lamb of the University of Chicago says that supernovae literally kick themselves out of the galactic disk, ricocheting off any other matter in the vicinity, exploding repeatedly and catastrophically as they do so.[16] Neutron stars were like cosmic land mines—vulnerable to anything approaching in their vicinity that could trigger them off and cause gamma rays to zap halfway around the universe. The emitted energy could be calculated in billions of electronvolts.

It is the astonishing distances gamma rays travel without appearing to lose any of their lethalness that is not only puzzling but extremely alarming. For the first time GRBs were implicated in the malfunction of spacecraft. In December 1997 an Italian-Dutch satellite called *Beppo-Sax* detected a flash of energy that shook the world of cosmology.[17] It lasted only a fifth of a second, yet held the record for the most intense burst ever detected by scientists.[18] The visible afterglow—largely of EM and x-rays and spotted by the Hubble space telescope—could only have been GRBs. Satellite findings of the event were decoded by astronomers at Kitt Peak Laboratory in Arizona, and at Hawaii.

There existed, in addition, a less violent but still pernicious kind of gamma ray. Physicists Robert Duncan and Christopher Thompson wrote in 1992 of SGRs—"soft gamma rays." They were emitted from magnetars (or magnetostars). These were akin to a neutron star, but instead of shattering, massively intense magnetic fields seeped out through their fantastically hot iron crusts. The magnetostar theory resurfaced in 1998 when for five minutes on 27 August the Earth's upper atmosphere was electrified by a burst of both gamma rays and x-rays. The rays came from a star called SGR1900+14 - some 20,000 light years away. They saturated the detectors of two spacecraft and disabled a third.[19]

Then there are the plasma clouds—great balls of high-speed gas and matter weighing an average of 10 billion tons—that periodically speed toward Earth at 2 million mph.[20] According to Glennys Farrar of Rutgers University, New Jersey, they strengthen the theory that there may be a new and menacing subatomic particle in existence.[21]

STORMS ON THE SUN

Some scientists are worried that as technologies change, new vulnerabilities to solar events will crop up. Radio receivers are experiencing more solar bursts than they used to, bursts that could eventually drown out radio networks. Satellite explosions also seem to correlate with 11-year sunspot cycles, as was alleged by scientists at the Institute for Astronomy at the Russian Academy of Science.[22]

When a magnetic storm hits Earth, power lines in large-scale electrical apparatus that act as conduits cannot cope with the massive EM fluctuations that are generated. The disastrous impact that solar storms can have on modern society was illustrated when the last but one big sunspot cycle took place in 1989.

Quebec suffered a massive power supply failure that covered huge areas of the province for four days and affected 6 million people. A voltage regulator had shut down on one of the main lines that run from the La Grande hydroelectric complex. Then four circuit breakers tripped, triggering a cascade of broken electrical connections around the region, cutting off the rest of Hydro-Quebec's generators. Power companies are vulnerable because these "geomagnetically induced currents" (GICs) catastrophically affect sensitive components such as transformers, which—as their name implies—change one level of voltage supply into another within specific discrete frequencies. If the power system is not shut down quickly enough, fluctuating voltages can rapidly pass through neighboring transformers like a high-power electronic virus, causing the entire system to implode. In all, it had taken less than 90 seconds for the entire La Grande grid to "go down," although power was restored in some parts of Quebec within nine hours. But the cost to the power station was $10 million, and probably twice that for Quebec's customers as a whole.[23]

Recall that Canada is near the North Pole, and it is significant that Sweden also experienced power cuts in the earlier 1989 solar storm. And Canada, like all developed countries, is investing heavily in computer technology—more so in fact because its 25 million population is spread out over an enormous geographic area. Its electronic infrastructure supplies phone and TV links all the way from Labrador to Vancouver. Further south a New Jersey power station burned out and had to be rebuilt at a cost of several million dollars. This was said to be due to a "giant plasma bubble" that had burst on the sun— in other words, the power failure was also the product of the 1989 sunspot cycle.[24]

The 1989 power blackout hinted at other hazards: during a geomagnetic disturbance, fluctuations in Earth's field can induce rogue electrical currents in Earth's surface. In poor-conducting regions like Quebec, which sits on a shield of igneous rock, the currents gladly surge through the power lines instead. The greater the distance the current has to travel—La Grande is 600 miles from Montreal—the greater the voltage difference induced by the magnetic fluctuations, and hence the stronger the current. Fortunately the additional currents add barely a few hundred kilowatts to the tens of thousands already coursing through the grid. Yet problems can arise because of the fact that most magnetic currents are direct, whereas grid currents are invariably alternating. This upsets voltage regulators and trips circuit breakers. And linking up grids into huge regional ones adds to the vulnerability.

Five years later, in 1994, Canada was again the victim of another plasma storm or "cloud" that disrupted large parts of the country's TV and electronic infrastructure. There were also many power "outages" in the U.S. West in 1996.[25] If the blackout had occurred during a heatwave or a cold snap, there could have been a heavy human toll.

However, it is not just large-scale electrical installations that are at risk. Electronics engineers from Texas Instruments and Intel in the United States and ST Microelectronics in France warned at a conference in Washington in October 1998 that the transistors in microchips are becoming so small that the background radiation of alpha particles and the like could easily corrupt the data they hold.

Most personal computers in the shops are powered by microchips with transistors now down to no more than 250 nanometers across, and fast heading for the extremely vulnerable 180 nanometer size. "What is currently an intermittent problem could easily become a much bigger one," says Gehan Armaratunga of Cambridge University. Electrical charges will be so tiny that each bit will be more susceptible to corruption by incoming radiation. Even the lead solder in the printed circuit naturally radiates alpha particles, says Alan Hales, an engineer at Texas Instruments in Dallas. Dispensing with lead is no help since a host of other materials in chip making, such as silica molds and phosphoric acid used for etching, are also natural alpha-emitters.

Systems could crash with disastrous results—literally, in the case of aircraft. What random zipping electrons can do to electronic equipment was hinted at by the oft-mentioned press stories about mobile phones—even when simply on stand-by—putting passenger aircraft in danger. In one instance in August 1999 a Chinese plane en route to Beijing from Chengdu in central China drifted 30 degrees off course because a mobile phone was not switched off.[26] An onboard mobile phone is believed to have killed 101 people in December 1998 when a Tahi Airways jet crashed while approaching Surfat Thani airport in Southern Thailand. So seriously do airlines take the issue that in June 1999 one man was convicted in an English court of "recklessly and negligently acting in a manner likely to endanger an aircraft" by refusing to switch off his mobile.[27]

It is worth mentioning, in fairness, that a direct link with navigational instrument failure has never been proven. Cellphones emit signals in ranges of around 400, 800, or 1,800 MHz, and no important piece of aircraft equipment operates at those frequencies. In one study conducted for Airbus Industries of Virginia it was virtually proven that cellphones have no impact on aircraft electronics.[28]

However a report released in 2000 by Britain's Civil Aviation Authority (CAA) said that perhaps older navigational equipment such as fuel level indicators, for instance, dating back to the 1980s, might be impaired. But some critics point to the fact that laptop and computer games, which are not banned on aircraft, are more of a threat to navigational safety because they can operate at a higher frequency range, even though they do not "transmit" as a cellphone does. An air crew on one flight found that the autopilot was being disconnected, and traced the fault to a passenger's portable computer.[29]

The CAA was more definite on the risks of mobile phones because of the transmitting nature involved. In tests aboard two airliners it confirmed that cellphone radiation itself interferes with flight-critical electronics. A team from CAA's Safety Regulation Group in tests carried out early in 2000 on Boeing airliners found that the aircraft generated signals at frequencies ranging from 380 to 1,700 megahertz at various points within the aircraft, and found that "interference levels exceed demonstrated susceptibility levels for aircraft equipment approved against earlier standards."[30]

Again, the problem was with older equipment. A key factor the tests uncovered was the varying power output of mobiles that could lead to fast-changing interference patterns. As an aircraft climbs, the mobile signals increase in power, boosting the interference level at a critical time.

The issue becomes complicated because a satellite functions, like an aircraft, in a soup of electronic emissions created by its own electronics. But unlike an aircraft it has to cope not only with ground to module transmissions but with the EM static of outer space unshielded by any dense atoms of the Earth's atmosphere. A combination of onboard electronics plus sky static can cause unpredictable behavior by reinforcing other signals or creating unforeseen harmonics that disrupt systems.

In America airflight passengers can phone a hotline to find out whether solar radiation levels are higher than usual. It informs callers of possible health risks and at what times solar storms cause radiation peaks.[31] Indeed in June 2000 NASA claimed to have empirical proof that space weather causes polar auroras that can disrupt satellite communication systems.[32] This proof was in fact known nearly 40 years ago. In July 1962 a Thor missile carried a 1.4-megaton warhead beyond the atmosphere, above Johnston Island in the Pacific. The explosion was visible from 2,000 kilometers away and the Electromagnetic Pulse (EMP) blast severely disrupted the street lighting system on several Hawaiian islands.

The event was, however, kept strictly secret at the time, and very little is known about it even to this day. But it is now known that it was called "Starfish Prime," part of a larger series of explosive tests done in space and beneath the oceans known as Operation Dominic.[33] In fact, Roger Grismore of the California Polytechnic State University, referred to earlier, blamed the uranium particles found on *Mir* on Starfish Prime. The particles from this event created an artificial aurora as they interacted with the lines of Earth's magnetic field. The aurora was visible for seven minutes. The high energy electrons that were created circulated through the Van Allen belt for months afterward, causing damage to early satellites—the first proof that some kind of antisatellite weapon could be developed by the Americans.

In the meantime, according to Joe Gurman of the NASA Goddard Space Flight Center, "We know that there is no such thing as a quiet sun."[34] A particular form of solar radiation that worries scientists is described as a "wind." The solar wind is like an eternal flow of electrified matter that bombards Earth's outer limits at speeds of up to 500 miles per second. The sun's heat varies constantly across and inside itself. The temperature of the sun's outer atmosphere is an astounding 1 million degrees, much hotter than its actual surface—the photosphere—which is about 20,000°F.

Magnetic storm predictions—similar to weather forecasts—involving electronic failures could come about one day in 2012, for instance, at the height of a new "sunspot cycle." Magnetic storms could produce winds of up to 500 mph in high-up cosmic places. In June 2000 NASA used data from its Polar Spacecraft and from the SOHO Solar and Heliospheric Observatory, launched back in 1996, to make early predictions. This means that given a few days notice, satellite operators could reschedule the trajectories of their spacecraft, and power companies could redistribute their loads.

Press and media weather forecasts now regularly give a "UV radiation index" on a scale of 1 to 10, indicating the amount of ultraviolet radiation expected in each area of the country during the day, with warnings to readers about the possibility of skin cancers.

Occasionally a "coronal hole" blasts out its own extra bit of solar wind, creating vast disturbances in near space, which is seething with innumerable tangled fields that surround the solar planets. In addition coronal mass ejections (CMEs, which used to be known as solar flares) take place, which have temperatures of more than 1m°C, according to Don Smart of the USAF Phillips Laboratory in Massachusetts. The largest, he says, carry energy enough to boil a lake several thousand times the size of the Caspian Sea.[35] However it is difficult to see all the CMEs heading toward us because most observing instruments need a black disc to obscure the bulk of the sun, so that only the corona can be observed. Further, over 60 percent of predicted storms never materialize.

SATELLITES AT RISK

The vulnerability of our hi-tech society to EM storms from space is closely connected to satellites. As mobile phones hardly existed at the time of the solar peak in 1989, JoAnn Joselyn of the National Oceanic and Atmospheric Administration (NOAA) told the national meeting of the American Astronomical Society earlier in 1999 that "the explosion in technology is intersecting with an extremely disturbed space environment." One CME was said to have knocked out an AT&T satellite in January 1997.[36] Along with cost-cutting and the consequent reduction of engine time testing and the need to carry heavier payloads, many specialists say it is the computer systems that are often to blame. But it is still not clear whether such failures are due to cosmic rays.

NASA warned in 1999 that its Deep Space Network, which tracks U.S. satellites, will be crippled by data pouring in from its missions to Mars.[37] *Deep Space-1* was launched earlier in October 1998 on a voyage to a distant asteroid named 1992 KD. It also flew by asteroid Braille in July 1999, and later took pictures of Mars. This $152 million probe is trying out new technologies on the way, and relying on its novel solar-powered xenon ion engine and its own star map for guidance. It was programmed to have to think for itself for about at least a week of the journey.

If the slightest error is introduced into *Deep Space*-1 in its calculations, it will rapidly become confused. In fact, its star-tracking camera did once fail, leaving the probe unable to orientate itself. Autonomous navigation software is no safeguard against this happening. Small changes in the temperature and pressure of the ions can vary the engine's thrust from one moment to the next. But it could easily be the sun storms themselves that can produce random power surges that can affect the ionic pressure. The French National Institute for Research in Computing Science (INRIA) came up with a "run-time error," meaning a kind of software fault. This fault carries a high risk, especially involving redundant code fragments—up to 90,000 in one Ariane-5 that exploded while in space.

Although the U.S. military is reticent about the issue, the U.S. Department of Defense has estimated that disruptions to government satellites from adverse "space weather" costs about $100 million a year.[38] The military could lose control of their satellites of the Missile Warning surveillance kind mentioned in Chapter 10—they could "go blind." Even in placid years, about 150 malfunctions occur. Defense publications suggest that several, or many, U.S. military satellites have gone out of control for sun storm or space weather reasons.

Dr. Dan Baker of the Space Science Institute in Boulder has pointed to the dangers of temporarily losing track of space junk, or confusing the effects of sun storms with microscopic particles of space junk. The NORAD observation post in Cheyenne Mountain in Colorado, although it has steel doors, can do little to protect its tracking equipment against sun storms. Baker says that if you lose contact with any satellite, military or not, "[then] if another object then comes into view, you don't know whether it is one of the 800 satellites or some new object, which could be a threat."[39] He warns that if a hostile state were intent on military mischief, it might take advantage of the disabling effects of a sun storm. For example, during the March 1989 sun storm there was a period of ten days when military observers were unable to identify over 100 space objects.

THE HAZARDS TO SATELLITE DIRECTION FINDERS

Interference in the ionosphere could also affect the GPS network, operated through the NAVSTAR flotilla controlled by the U.S. Air Force. If major airlines use GPS more in the future, sun storms could present a hazard to travellers, or at least cause diversions or delays. As long as you can get your signals from at least four spacecraft, you can know where you are within 100 meters, including altitude and momentum. With access to a second coded military GPS receiver, this can be improved hundreds of times over, down to centimeters and millionths of a second. Military GPS receivers are more accurate than those available to the general public because they can access coded transmissions on a second satellite frequency. But even if the GPS satellites are robust enough to weather a sun storm, their signals can still be distorted. In the 1989 storm, there were reports of "major errors" even in the sophisticated receivers used by the U.S. military.[40]

Dave Pike, a scientist working with the *SOHO* satellite, said that even an hour's warning could help to reduce damage to electronic circuits: "Power circuits are more prone to damage if they are already carrying a heavy current," he said. "If utilities can reduce the load, there is less chance that the surge of current from solar activity will blow the circuits. The same applies to satellites."[41]

Take, for example, the situation that occurred in the spring of 1996. Two Canadian Telesat satellites, *Anik E1* and *Anik E2*, succumbed to solar storms with widespread knock-on effects. For nearly five years, at a cost of $220 million each, they had circled the Earth without a hitch. Then, until users could switch to backup systems, the transmission of all sorts of critical data—credit-card transactions, electronic paging, TV, radio broadcasts, internet link-ups, and even newspaper layouts—stopped.[42] Some scientists said a spark short-circuited the satellite's solar-powered panels and several dozen of its radio relays. *Anik E2* was more seriously damaged. According to Peg Shea and Don Smart of the U.S. Phillips Laboratory in Massachusetts, rogue electric currents commanded the satellites to turn their solar panels away from the sun. E2's batteries drained away, and it floated out of control for six months.[43]

Spacecraft that navigate by the stars (using equipment known as "star trackers") can be disorientated by high energy solar particles. William Ailor, of the Center for Orbiting and Re-Entering Debris Studies at the Aerospace Corporation, spoke of similar dangers from space plasma before a congressional subcommittee studying the likely effects from the Leonid meteor shower.[44] In the past four years, says Chris Kunstadter of U.S. Aviation Underwriters, space losses may have exceeded $550 million. "Solar storms can cause millions of pounds worth of damage to satellites. But if you know the expected time of arrival of one of these clouds you can shut your satellites down," said Clive Dye of DERA.[45] As the particles collide continuously with others in space, they emit light that concentrates into the blue Cerenkov radiation flashes, the optical equivalent of sonic booms. Software on the satellite's star tracker can filter out these flashes to avoid their being mistaken for distant stars.

SPACE WEATHER AND COMMUNICATION FAILURES

Space weather can affect satellites in several ways. During bad spells, particles from radiation "belts" surrounding Earth can affect high-orbiting satellites because they become more energetic, so that they charge up spacecraft surfaces more, causing sparks that can damage the surfaces and disrupt circuits. If more electrons arrive than can leak away, they build up a powerful electrical field in the spacecraft, a process known as "dielectric charging." After a threshold is reached, it discharges in a massive burst to burn out sensitive components. What struck the *Anik* satellites was probably something more akin to a bolt of lightning. Caught in the plasma cloud, the outer surfaces of the satellites would have been suffused with a heavy static electrical charge. This would eventually have earthed itself, and a huge spark would have flashed through the crafts' circuitry causing "phantom commands" or the burning out of some component. It took engineers five months to get *Anik E1* going again. If Canada had lost the two comsats for good, the cost to Telesat would have been crippling. Satellites take a long time to design and build, and the company in the meantime still has to serve its customers.

Similarly, AT&T's satellite, *Telstar 401*, which was relaying TV broadcasts and phone data, was lost on 17 January 1997. This disaster could hardly have come at a worse time because the company was on the verge of selling three satellites, including the 401, to another company, Loral. But when *Telstar* went out of action, many TV screens turned to "snow."[46] The similar failure of the Hughes Electronics Corporation's comsat *Galaxy IV* in May 1998 was also blamed on solar storms.[47]

One spacecraft, operated by PanAmSat, of which 80 percent is owned by Hughes Electronics, had been hit by sky static in May 1998 and had spun out of control, interrupting broadcast and cable TV transmissions, including all 600 stations of the U.S. National Public Radio service, as well as radio stations in Denver, Minnesota, Chicago, Seattle, Portland, and Los Angeles. The satellite was in geosynchronous orbit 22,000 miles above Kansas. *USA Today* called it "the biggest telecommunications failure in recent years," adding that the breakdown had "wiped out pager traffic, halted credit card transactions and knocked TV and radio stations off the air." The cause was the failure of an on-board navigational computer and its backup, although the company admitted it knew no reason why the computer failed. Not only that, but four months later the x-ray astronomy satellite *ROSAT* was rendered useless after it drifted out of control and pointed itself at the sun, damaging its high-resolution imager, its last working detector.

Another violent solar storm took place on 14 July 2000 and was monitored by the *GOES-8* satellite, which relayed its findings back to the Space Environment Center in Boulder, Colorado. The satellite registered a sharp jump in the intensity of x-rays emanating from the sun. In fact, the flare temporarily blinded the instruments of another observation satellite, the *SOHO*, and destroyed another satellite, the Advanced Satellite for Cosmology and Physics, and an x-ray observatory launched in 1993.[48]

Aircraft and satellite avionics face the worst problems as microchips shrink, since they are literally closer to the cosmic radiation. "One proton zipping through the wrong piece of electronics can render a satellite useless," said David Hathaway, a NASA solar physicist. Electrons can get knocked out of their shells by fast-moving neutrons that can accumulate in the crystal lattices of microchips, where they build up a corrupting charge.

Alan Hales of Texas Instruments in Dallas says only ten feet of concrete around each computer would solve the problem.[49] A group of researchers at the Ericsson Saab Avionic in Linkoping have been investigating cosmic rays that can have an impact on electronic components that contain static random access memory (SRAAM). They found that the binary code of 1s and 0s can be altered. Rays with an energy of more than five megaelectron volts can change a 0 into a 1 or vice versa.[50]

The effects of cosmic rays were observed during the 1996 Euromir mission, when two IBM laptops were tested for use in space. It was found that cosmic rays caused memory errors once every nine hours. A SRAAM device was tested aboard a Scandinavian Airlines long-haul jet for 1,000 hours above 10 kilometers. On average it found that a piece of information on the device was changed by a cosmic ray every couple of hours. Karin Johansson, a research engineer at Ericsson Saab Avionics, says the tight integration of avionics systems can trigger a great raft of problems, resulting in a domino effect.

One solution would be to double the number of transistors used to store each bit of data, but this would be expensive and would naturally build in component redundancy. Only Boeing 777 and Saab aircraft use software checks to guard against errors induced by cosmic rays. For those satellites in lower orbits it is atmospheric drag that is the problem. This is because UV light and magnetic storms also heat up the atmosphere. This in turn means the density of rarefied gases increases. The added drag for a satellite 400 kilometers up can be 20-fold, enough pull it into a lower orbit. NASA's *Skylab* was one victim of atmospheric heating—it plunged back to Earth in July 1979 after a period of high solar activity. Duncan Steel suggests this might be a good thing, bringing space junk down all the sooner.[51]

EARTH'S TRICKY MAGNETIC FIELD

There is one cause for optimism: that the worst excesses of sun storms can be mitigated—by Earth itself. The Earth's magnetosphere generally shrugs off the sun's own magnetic field lines, but it looks as if it is an uneven battle. The magnetosphere, after all, is only about half a mile thick. Earth's magnetosphere is a long, teardrop-shaped region that surrounds its near space. Daniel Baker of the Lab for Atmospheric and Space Physics at the University of Colorado, has compared it to a classic chaotic system: a plasma flowing in from the sun collects in its tail in a drip-drip manner until it reaches a critical mass. Magnetic lines snap back toward Earth, generating intense electric currents.[52] Ironically the repellent effect of Earth's magnetic field was discovered by the first American satellite, *Explorer-1*, in the 1950s. Particles were found to be trapped, some by a field that extends some 30,000 miles up.

The result of this event can be truly spectacular to those living beyond the Arctic circle or in the region of the Southern Ocean. The aurora borealis is one of nature's most beautiful and spectacular sights. It radiates across the color spectrum, sometimes appearing diffuse and at other times sharply defined. It appears as a series of curtains that glow dramatically, extending across the sky of both polar regions in a series of veils, bands, and arcs. The field is weaker at the poles and so allows some of the charged particles to enter Earth's atmosphere. Energy unleashed by the solar wind excites atoms of nitrogen and oxygen to glow much like phosphorus in a TV screen.

NASA, in June 2000, showed that a number of polar spacecraft travelled directly through a number of "tears," or rips, in the Earth's field. One of *Solar*'s tasks was to see how the solar wind interacted with the magnetosphere. Many scientists, including Jack Scudder of the University of Iowa and Jim Drake, a theoretical physicist at the University of Maryland in College Park, were surprised at these findings.[53]

The impact of solar bursts on Earth depends essentially on the current state of geomagnetic play—whether the Earth's field is increasing or decreasing, and in what way, and in which hemispheres. Yet there have been many false alarms, as in the case of the flare that occurred in January 1997, which attracted the attention of a disappointed media that accused the flare of being "oversold."[54]

Much depends on polarity, and there is no way yet of predicting this. Unfortunately the field tends to vary from one part of the world to another and from one epoch to another, and its polarity (north or south) depends on a complex relationship with the sun's own polarity. A pair of spots become magnetically polarized, with each one having a north or south polarity. Whatever the north polar or south polar arrangement, the situation in the opposite hemisphere of the sun is curiously reversed. Not only that, but after the 11-year cycle has ended, all the matching polarities are reversed for the next 11-year cycle.[55]

Sometimes the storm can be repelled and sometimes—if the two fields are opposites—they can reinforce each other. For this reason it is difficult to predict geomagnetic storms. One form of electromagnetic activity can increase on the sun relative to the rest.

The USAF is now running a computer system at its 50th Weather Squadron near Colorado Springs to predict when and where electric storms will occur. Developed by scientists at the Aerospace Corp in Los Angeles, California, it is called the Magnetospheric Specification Model.[56] There is a need to predict solar bursts and nudge satellites out of the way.

Over the 11-year sunspot cycle, the strength of the field varies enormously. Some spots can last months, others only a few hours. They emerge alongside the magnetic lines of force with their own fields, at least 1,000 times stronger than both the field of the Earth and the rest of the sun. During a "solar minimum," to complicate matters (when in theory there are supposed to be few sunspots), big blobs of energy called coronal holes come from openings in the sun's magnetic field. "Gas spews out of these holes," explains University of Colorado space physicist Daniel Baker, "like water from a fire hose."[57]

By chance a group of scientists, members of the International Solar

Terrestrial Physics Program, were at a NASA meeting studying the solar flares revealed by *SOHO* at the time *Telstar 401* was knocked out in 1997. Dr. John Dudeney, of the British Antarctic Survey, had analyzed the *SOHO* data and confirmed that the cloud was heading toward Earth in that year. But the storm had caused a lot of radio interference, which had made it difficult for the Antarctic scientists to contact their own bases—which were separated by hundreds of miles—and to operate aircraft communication and navigation systems. At the same time, the *GOES-8* satellite also malfunctioned. Many had no doubt it was the plasma cloud that crippled *Telstar 401*, although not all were agreed. Further, late in January 2000, even NASA's *Advanced Composition Explorer* (*ACE*) found itself under assault from solar particles connected with the solar maximum predicted for 2000–2001. In the space-environment control room at NOAA headquarters at Boulder, alarms sounded. "All of a sudden a blast wave of solar wind showed up at the *ACE* spacecraft as dense as any we've seen . . . thirty minutes later the Earth's magnetic field got hit hard," said Joe Hirman.[58]

For the time being, the impact of CMEs can be unpredictable, like not knowing what kind of weapon your enemy is using against you. Dr. James Baker, head of NOAA, told a Radio-4 BBC journalist in November 1999 that fast-moving solar particles can take 20 minutes to reach Earth and its satellites, too short for a warning to be issued. But *ACE* can provide an hour's warning of slower-moving particles from CMEs.[59]

KEEPING AN EYE ON THE SUN

Forecasters at the Space Environment Center in Boulder keep a constant eye on the sun every midday and at least try to predict whether a geomagnetic storm might be in the offing. Some scientists have had considerable success at forecasting when the CMEs will strike Earth. Sometimes the CMEs can be detected by simple changes in radio signals from observation satellites in space. Montana State University scientists spotted S-shaped patterns appearing on the sun's surface a few days before a coronal eruption occurred. If the pattern crosses a sunspot area, an eruption is virtually certain.[60]

At the International Solar-Terrestrial Physics Program (ISTP) a mass ejection on 7 January 1997 was detected just two hours later by a range of observing satellites, including the Solar and Heliospheric Observatory (SOHO), and by the *Wind* satellite, which monitors the solar wind from an orbit shaped like a figure eight. Another space-craft, the *Geotail*, also observes the zone affected by Earth's magnetic field. The ISTP physicists predicted that the solar eruption would reach Earth three days later, which it did, and it triggered a powerful magnetic storm over Antarctica, jamming important radio commu-nications with Britain's Halley Research Station.[61]

Yohkoh, a joint Japanese-American satellite with British participa-tion, was another solar predictor. "This is the first step in providing future warnings on these events and may help mitigate the power blackouts and satellite degradation that may be caused," said Lord Sainsbury, Britain's science minister, who took a close interest in the *Yohkoh* project.

There is also NASA's *Image* spacecraft, with an antenna longer than the Empire State Building is tall, which also checks into the solar wind. The *Image* "will revolutionize our study of the magnetosphere," says principal investigator James Burch of NASA.[62] In November 1999 the NOAA inaugurated new space weather scales, the cosmic equivalent of the Beaufort wind scale. NOAA has five levels for as-sessing the likely impact of a solar storm, ranging from five, the most extreme, to one at which no effects are expected.[63]

Further, a new international high-frequency radar network using phased-array techniques is being installed in Alaska and is known as SuperDARN. It will also measure global-scale magnetospheric con-vection characteristics by observing plasma motion in the Earth's up-per atmosphere.

In conclusion, space scientists are more worried about the effects of solar and space radiation on spacecraft electronics than most of the other threats coming from deep space. Can the electronic guid-ance and communication technologies of satellites and space stations be sufficiently protected by this form of "space weather"?

Ashot Chilingarian runs a cosmic ray observatory at the top of Mount Aragats in Armenia, a leftover from the Cold War. It is the leading laboratory of its kind in the world and can offer reliable early warnings of severe solar storms that can cause billions of dollars' worth of damage to satellites and power systems. It has an array of detectors covering hundreds of square meters to measure the much smaller fluxes of high-energy particles that come ahead of the lower energy particles that NASA's science satellites can detect. Using com-puters, the Armenians can provide the most reliable forecasts of such storms, and are seeking commercial contracts.[64]

However, as we have seen, even the massive steel doors at NO-

RAD's Cheyenne Mountain site can do little to repel the effects of solar storms. But if the static is a phenomenon associated with, for example, meteoroid storms, then the answer is probably yes. But whether cladding and shielding alone will be sufficient in complex orbiting computers awash in their own electronic soup is still unknown. One can understand why not only western military powers are concerned about the threat from natural cosmic phenomena. Historically the space environment has been a subject of interest only to the science fraternity. But as society becomes increasingly dependent on electronic information largely fed by satellites, changes in the "space weather" that now include damaging radiation and zapping particles have a wider import: they are of direct concern to defense institutions and satellite manufacturers.

As we shall see in the following two chapters, there are many ingenious schemes afoot that aim to come to terms with this threat.

CHAPTER 10

Protect Our Spacecraft

Concepts of "unimpeachable airspace" have always been important to the United States Air Force (USAF) and its associated defense bureaucracies. All of them acknowledge, via surveillance intelligence, the threat to America. Missiles plunging through the sovereign airspace of a nation can have obvious defense and geopolitical implications, as we saw in Chapter 5. These events can produce fiery displays, are often accompanied by smoke trails, and can give the impression of either a missile attack or a collision between two aircraft.

A group of NASA and Lockheed-Martin scientists said the first NASA attempt to calculate the probability of a collision between debris and a manned spacecraft occurred in 1966 at the time of the *Gemini-8* mission.[1] Dr. Sasumu Toda pointed to another early warning of satellite collision, which was made in Japan in 1971 by M. Nagatomo and his colleagues at the Institute of Space and Astronautical Science.[2]

The Japanese National Space Development Agency (Nasda) focuses especially on space debris, and has done so since 1994. But, as revealed recently, it is assisted by most of the other Japanese space and aerospace or astronomical consortiums, such as the Japan Space Forum, the National Astronomical Observatory, and the Japanese

Spaceguard Association (JSGA), amounting to about a dozen in all, most of which are based in Tokyo.

The debris issue is, as we have seen, regarded as both a military and a space industry problem. The recognition of the hazards of space debris came as early as 1988. Although natural processes such as the expansion and contraction of Earth's atmosphere cleanse hundreds of tons of debris from low orbits annually, there was pessimism about what could be done at higher orbits.

Nevertheless, the ESA issued a safety policy in that year, which was later incorporated into the policy document known as the European Cooperation for Space Standardization (ECSS). The Phillips Laboratory began its debris modeling and mitigation analysis in 1994, and draft regulations were written up by the laboratory for DoD to include the aims of U.S. space policy and to genuflect to Air Force guidelines.

A year later the U.S. National Research Council warned in a report, which included experts from the U.S., Japan, Canada, Russia, and Germany, that if nothing was done about space debris, inner space could become so clogged with high-speed orbiting flotsam that it would become a "death zone," presenting a potentially lethal hazard to spacecraft of all types, manned and unmanned. In the following year, 1996, the ESA declared that mitigation policies must include the removal of rocket launchers after mission completions.

In 1991 the International Astronomical Union formed a Working Group on Near Earth Objects (NEOs). A year later a subcommittee was set up to investigate "asteroid hazards" in conjunction with the International Institute of Problems of Asteroid Hazards (IIPAH). It was the continued fear of an impact, or the threat of serious disruption to spacecraft flights, that goaded the U.S. Defense Department to issue a document and present it to the USAF in 1996.[3] This document in turn resulted in a directive to the chiefs of staff regarding the technologies they would have to use to maintain their domination of the skies. The document, in effect, deplored the "current lack of adequate means of detection, command and control . . . in terms of the courses of action in the event of a likely impact by an ECO (Earth-crossing-object)." A NASA-DoD working group was established a year later, and a U.S. Government Orbital Debris Mitigation Standard Practices was set up. An Orbital Debris Workshop for Industry was set up a year after that.

Another reconvened joint team of American space scientists from

NASA, the Marshall Space Flight Center in Huntsville, Alabama, the Jet Propulsion Laboratory (JPL) at Pasadena, and the Langley Research Center in Virginia, pointed out the strategic and scientific importance of comsats and elint satellites, which must be protected from plasma and particulate sky static. Mitigation studies were done at the USAF Space Warfare Center at Shriever AFB near Colorado Springs, and also at Kirkland AFB in Albuquerque, New Mexico. Important debris studies were conducted by private concerns like ITT Industries, Lockheed Martin, Teledyne Brown Engineering, Kaman Sciences Corporation, Boeing North American, Batelle, and others.

Increasingly, mission planners now consider the risk of impacts to the shuttle and plan its orientation accordingly. Occasionally the shuttle executes an evasive maneuver, as we have seen, to dodge large, live satellites. Often the most sensitive part of the spacecraft surfaces are—as demonstrated in the 1998 Leonid storm—pointed away from perceived threatening directions.

A paper by Dr. Nigel Holloway of the Atomic Weapons Establishment in Britain reviewed the risk of NEO impacts in the context of the "tolerability of risk," a concept introduced by the U.K. safety regulators of the Health and Safety Executive, following the Sizewell B public inquiry.

Many scientific institutions and observatories take part in NEO research. The detector GORID was successfully launched on board the Russian *Express-2* comsat in September 1996 with a Proton rocket from Baikonur, mainly to check on asteroid debris.[4] It is expected to last at least seven years in orbit, according to a joint team of experts from Holland, Germany, and Russia headed by Dr. G. Drolshagen of the Dutch space debris research organization, ESTEC (the European Space Research and Technology Centre). They said that during the first 118 days of operation, several hundred events were recorded, "many showing signs of real impacts."[5] Now spacecraft protection has gone hand in hand with new designs based on new guidelines and "basic technologies" derived from a database that includes results of hypervelocity impacts and state-of-the-art space vehicle protection systems.[6] Hypervelocity impact research bears on many areas of modern physics, geoscience, and astronomy itself. New types of accelerators have been made to emulate projectile impacts and include electrostatic and EM-propulsion and explosion propulsion techniques. Studies have been done, as we have seen, with "light gas guns" that can fire projectiles at a velocity of up to 3 kilometers per second.

THE FLYPAPER EXPERIMENTS

Just a month before the remains of the *Mir* crashed into the Pacific Ocean, it threw up one last puzzle. Radioactive elements, as we have seen, of uranium were discovered on one of its instrument covers. This was odd. It could have been the result of fallout from earlier nuclear weapons tests conducted some 30 years earlier. This was the conclusion of scientists from California Polytechnic State University in San Luis Obispo. Perhaps, they surmised, it could have come from nuclear powered satellites. Some of these craft used depleted uranium as ballast. Two uranium-fueled *Kosmos* satellites burned up around 20 years ago, pointed out John Zarnecki of the Open University in Milton Keynes, England.

But what was important about this discovery was that the principles behind the method of detection of the uranium could also be utilized for the protection of other satellites and space stations. Roger Grismore and his Californian researchers came across the uranium by accident after a small mitten-shaped space blanket was placed over a glass instrument on the outside of *Mir*. It consisted of ten thin layers of aluminum and polyester. It was made to protect the craft from solar radiation and tiny meteorites. It was removed in August 1995 and kept in quarantine on Earth for 16 months before the researchers took a detailed look at it. Spectrometers revealed gamma radiation produced by telltale isotopes of gamma rays that could only have come from uranium-238.[7]

The number of debris-monitoring craft launched specifically on science missions—all acting like mechanical flypapers in space—now number at least 30. A Long Duration Exposure Facility (LDEF) was launched by NASA recently to complement several LDEFs sent earlier in the 1980s.[8] It is a kind of deliberate target: scientists want to see what would happen to it after six years in orbit. The LDEF were purposely designed to experience natural space debris, and the latest was struck by tens of thousands of artificial shards, as well as natural meteoroids and space debris, between the years 1984–1990.[9] NASA found that a piece of aluminum debris less than 1 millimeter had smashed through an aluminum wall 2.5 centimeters thick. Evidence was also found that satellites were being literally sand-blasted by spacedust particles: enough, according to NASA, "to degrade mission performance or cause mission denial."[10] According to John Stark of London University and Richard Crowther of DERA, the LDEF craft revealed, in a report submitted to the Second European Conference on Space Debris, that some 34,000 impact features were noted, of which nearly a quarter were artificial.[11]

In other cases, existing spacecraft were fitted with experimental dust detectors. During the STS-66 space dust detector mission, the shuttle *Orbiter* windows were exposed continuously to "ram direction particles."[12] In fact, the pitted windows of the returning shuttles first hinted at the existence of a large population of small paint particles in orbit.[13] The Eureca detector attached to the Hubble telescope flights also showed similar effects. In February 2002 it was revealed that the Hubble space telescope had suffered over 570 hits during over 10 years of operation. Collisions with millimeter-scale particles at speeds reaching 36,000 km/h have left centimeter-sized craters in it.[14] Dust detectors were also fitted to the *Ulysses* and *Galileo* spacecraft and to the *Giotto*, *Vega*, *Hitan*, and on the *Mir*. Other detecting spacecraft have included the *European Retrievable Carrier* and Japan's *Space Flyer Unit*.

The most well known space dust detector is SPADUS, devised by scientists at the University of Chicago in its USAF-sponsored Space Test Program.[15] This was put into orbit aboard a military satellite (the USAF Advanced Research and Global Observation Satellite). It is able to distinguish between natural and man-made particles, and checks on their size spectrum and geocentric trajectories. Up to the end of 2000, SPADUS recorded 327 impacts, and in late March 2000 the detection rate dramatically increased, although the source of the impacts was unknown.[16] It complements the similar dust-analyzing techniques fitted to the Argos Advanced Research and Global Observation satellite *P91-1*, sponsored by the USAF Space Test Program.

SPADUS uses two thin sensor layers stacked 20 centimeters apart. They are in effect an array of foils producing tiny electrical pulses whenever they are penetrated by specks of dust. The size of the pulse gives some hint as to the speed and size of the particle. Some particles are energetic enough to puncture and pass through the top layer into the second layer, and the speed is then even easier to work out (it acts like a car going through a radar speed trap).[17] After a year in orbit from 1998 to 1999, data from SPADUS revealed 159 impacts. Ten were energetic enough to penetrate the second layer of foil. Seven proved to be man-made and three were interplanetary in origin.

Another three-year mission, jointly run by the University of Chicago and Lockheed Martin Space Sciences Laboratory project at Palo Alto, California, together with the U.S. Naval Research Laboratory in Washington, will perform experiments similar to those made with SPADUS. In addition, as part of the EuroMir group of experiments (using *Mir* as a "flypaper"), payloads with surfaces to monitor the potential impacts hazards in LEO were installed on a platform called the European Science Exposure Facility (ESEF).[18]

A similar experiment using ultrasensitive foil layers to record electrical impulses was called the Stardust project. *Stardust* has a Large Area Momentum Sensor (LAMS) mounted on its front bumper shield. It was built by U.K. engineers at Kent University, funded by a grant from PPARFC, the U.K. Particle Physics and Astronomy Research Council, and launched by NASA from the Kennedy Space Center in Florida in February 1999. The space engine is to begin a 3-billion-mile journey during which it should achieve a first for space exploration by gathering material from beyond the sun and the moon. Its target is *Wild-2*, and in 2004, the craft will approach it at 14,000 mph to capture the tiny dust particles that make up the comet's tail, eventually returning them to Earth in January 2006.[19] Larger grains that penetrate the foil layers will be detected by a microphone fixed to a layer of Nextel™ cloth. At the rear of the shield is a set of microphones designed by the Kent scientists in collaboration with the University of Chicago to "listen" to the sound of dust particles as they strike the spacecraft.

A team of scientists, including John Stark from the Department of Engineering at Queen Mary College, London, and Richard Crowther from DERA, concentrate on the dangers of atomic oxygen (atox) impacts in space. These are often evident on the surfaces of returning spacecraft and arise partly from flaking paint particles caused by a "synergy" between atox and the "thermo-mechanical properties of surfaces."[20]

REPELLING SPACE JUNK

In January 1998 the U.S. government presented draft debris-mitigation standards to the aerospace industry for comment. Richard Crowther of DERA said that the UN had just come to the end of a three-year review on space debris, and that 1999 would be the year "we will have to decide how we are going to deal with it. We would expect in future that when people do use space they would leave the environment almost as they found it."[21]

Strategies were being cautiously formulated, initially being based on plain common sense and the precautionary principle. One idea would be to reduce the amount of debris dumped in space or, instead, simply to try to dodge the debris. Right now only the *ISS* and the space shuttle make a point of getting out of the way of junk, says Nick Johnson of NASA. They plan their collision-avoidance routines via a detailed computer model, but the calculations have been criticized as being too complex. Shuttle astronauts get a warning whenever a piece of debris will come anywhere near a 3 by 15.5 by 3-mile wide region near to the spacecraft. This then triggers an "alert box" on the craft. But they must definitely dodge anything heading into a 2.5 by 6 by 2.5-mile "maneuver box."[22]

The collision-avoidance model, however, can be simplified to just one equation, according to Russell Patera of the Aerospace Corporation in Los Angeles, by taking the influences of the Earth's gravity into account and ignoring other complexities such as other gravitational or cosmic influences. This will then reduce the calculation time to just seconds, he says.[23]

Larger objects can be tracked as they circle the Earth, and ground operators can maneuver satellites and spacecraft around them. Most launch nations already drain residual fuel from junked rocket stages, although dead rockets and satellites could be prevented from exploding and spewing shrapnel throughout space. Mission controllers can even use any extra fuel to swing a rocket into a fast-decaying orbit.

Unilateral action by the space powers sometimes complicates the issue. The Japanese are toying with the Orbital Maintenance System (OMS), which is a service satellite specifically designed to inspect and repair failed satellites, or to remove them from orbit.[24]

However space objects are protected by international law, and there are agreements not to use "anti-satellite weapons." This effectively excludes, in theory, attempts to remove a satellite, alive or "dead," from orbit. Unfortunately, since the end of the Cold War, international law has not kept up with the vast number of inactive objects now in space, as Lubos Perek, of the Czech Astronomical Institute, pointed out. Still, many countries may well not want other powers removing their satellites, which could harbor technological secrets.

But Perek says that existing space law can still be used for debris removal. He points to the Registration Convention, which comes under Article III of the Treaty of Outer Space, passed in the early 1960s. Article IV says that each spacefaring state shall furnish launching details and registration numbers to the Secretary-General of the UN for spacecraft that are no longer functional or transmitting.[25] But the UN Registry has gotten bogged down in bureaucratic and informational overload, and there is no legal requirement for information deadlines. Some launching powers submit launch data years after the event, and the space object population count soon becomes hopelessly out of date.

In the meantime, the Japanese are trying to limit the orbital life of their *H-II* second stages, as was done on 28 August 1994, when the craft was put into a "burn mode" so that it exhausted its fuel.[26] They have tightened up their separation devices and made strenuous efforts to minimize the orbital life of the spent upper stages. Simple economy measures were also suggested. NASA and NASDA recommended that new LEOs be deorbited.[27] Another idea is to push dead spacecraft upwards into a "graveyard orbit," which is at least 300 kilometers above the GEO zone. Walter Flury, in his report to the Second Conference, pointed out that *GEOS-2* was pushed into a graveyard orbit, as were *OTS-2* and *Meteosat-2* in 1991.[28] Eamonn Daly, who heads the Spacecraft Environment Section at ESA, says the answer is to reorient the spacecraft's orbits or to launch differently, which he admits is expensive.[29]

Another idea is to enable a satellite to deorbit itself at the end of its life with a piece of wire dangled from it—an electrodynamic tether, in effect, sometimes up to five kilometers long! In other words, Earth's magnetic field can be utilized to set up a current in the wire.

Electromagnetic space influences charge up the tether, and this in turn slows or disturbs the craft's orbit. At the same time, it can also save hundreds of kilograms worth of propellant that would be needed for re-booster rockets.

The process can also work in reverse: by passing a current through the tether, a satellite can increase its altitude. Indeed some inventors suggest that on a far more elaborate scale, spacecraft can even be launched to the inner planets using a combined technique of electrodynamic tethers, or "rotating-momentum transfer tethers," to hurl payloads into space. The payloads then use "sling-shot" techniques involving planetary gravitational fields to gain velocity momentum and complete their journeys.[30] It also obviates the wasteful need for a satellite to keep some of its fuel in reserve for deorbiting: fuel can amount to 20 percent of its mass. For a tether, the figure is only 2 percent of its mass. NASA tested the idea in 2000 by deploying a 15-kilometer tether from the upper stage of a rocket. One commercial concern, Tethers Unlimited, has a NASA contract to design satellite tethers.

One solution would be to get satellites to return to Earth to be reused (NASA's space shuttle is the only spacecraft currently being reused). Department of Defense researchers have developed a robot that can refuel and service America's spy satellites while they are in orbit. This could extend a satellite's life and obviate the need for it to drop out of orbit once its fuel was used.

The robot refueller was dubbed the Autonomous Space Transporter and Robotic Orbiter (ASTRO) and will shuttle back and forth between the spy satellite and fuel dumps stationed in holding orbits, says David Whelan, director of the tactical technology office of the Defense Advanced Research Projects Agency (DARPA).[31] "If an airplane runs out of fuel you don't throw it away," says Charles Miller of Constellation Services International in Dayton, Ohio. And yet this is what we do with satellites costing as much as $1 billion apiece, he says. DARPA has secured $5 million to begin designing ASTRO, and has commissioned contractors to build prototypes. By adding modular electronics systems, the robot could also be used to replace faulty or outdated on-board systems.

DEFEATING THE METEOROID STORMS

In regard to mitigating the effects of meteoroid storms, Intelsat said there were a number of precautions that could be taken: (1) Reschedule all station-keeping functions, (2) Disable the thrust mechanism and make the propulsion subsystem safe, and (3) Rotate the solar panels parallel to the direction of the storm, even though this option depends on the battery consumption and capacity.

Often in November when the storm is at its height, the direction of the sun is roughly perpendicular to the incoming track of the storm, and the solar arrays are generally pointing towards the sun to catch the maximum amount of energy.

The U.S. military paid NASA $17 million in 1997 to build a solar x-ray imager, which was flown aboard a weather satellite in 1999. NASA is also building Ace, the Advanced Composition Explorer, to give virtual fail-safe one-hour advance warnings of severe geomagnetic storms. Also at the peak of a November storm, SES could make slight adjustments to the arrays, and like Intelsat they will have technical staff on hand.[32]

At present the insulation cladding to protect satellites against heat and light damage is nonconducting and allows electrostatic charges from solar particles to build up on the satellite's skin, leading to arcing. Now Rui Resendes and colleagues at the University of Toronto have developed polyferrocene, which has iron atoms bound into its molecular structure.[33] This allows the charge to equalize around the surface of the satellite. "This is certainly a valid approach. Ultimately it will come down to cost," says Alan Tribble, an aerospace engineer with Rockwell Collins in Cedar Rapids, Iowa. Orbital Sciences Corporation, who designed and operated the Pegasus (which was destroyed in 1996) have redesigned the upper stage and implemented new preventive measures. NASA and NASDA have used bolt catchers and special tethers to limit the release of debris, and have jettisoned fuel on dead spacecraft to prevent explosions.[34]

Most satellites have highly customized software for communicating with Earth. Recently, Surrey Satellite Technology of Guildford, Surrey, was commissioned by NASA to see whether satellites can be controlled using cheaper civilian computer technology. The practice of "colocating" a number of satellites in the same orbits so that a single TV dish can pick up all signals is becoming too challenging for existing technologies.

Software improvements could however include getting rid of corrupt data and eliminating security breaches that could send uncontrolled commands to external spacecraft devices, all by using computer techniques. Martyn Thomas of Praxis Critical Systems in Bath, England, believes getting the coding and software right in the first place is more important than searching out old code fragments that should not be there.

Finding dangerous errors in mission-critical software programs before they are deployed for real is now being stepped up. The Polyspace Technologies Company of Grenoble deploys a computer bug-buster for the space industry.[35] INRIA, the French computer science institute, is also examining debugging techniques. Batteries can also be made safe to prevent static building up and making them accumulate more charge than they can hold. The batteries could be fully discharged and then short-circuited to prevent this, although virtually nobody currently does this, according to Nicholas Johnson. NASA's Meteoroid and Orbital Debris Technology Program (MOD-TP), to check into technologies to build tougher, survivable spacecraft, has been started up. Improved spacecraft designs and new types of materials could halt peeling paint as a source of debris.

Some speak of "robot cleaners" that could "pick up the pieces." Kumar Ramohalli at the University of Arizona began work in 1988 on a robotic space cleaner called ASPOD (Autonomous Space Processor for Orbital Debris). This would use a solar powered laser to shatter decommissioned satellites into pieces. Robotic arms would recover reusable parts such as solar panels, after which the ASPOD itself would be recovered after it had splashed down into the sea with its haul.[36]

ZAPPING SPACE JUNK

One recent solution was simply to zap space junk with a giant laser. NASA's studies of megawatt laser pulses on different types of debris have shown that a laser broom would be best at slowing it down or reflecting it: in effect, it would be able to deorbit debris. The studies, according to Jonathan Campbell of the Marshall Space Flight Center in Huntsville, Alabama, have been "very promising."[37]

However the problem of first detecting debris of that size from either space or the ground is admittedly the most difficult task, and renders the laser broom rather academic. One idea is for the *ISS* to release simulated debris equipped with the GPS so that its position can be monitored and "lit up" by the ground-based laser system.

Lasers in fact could vaporize debris rather than blasting it into further tiny pieces. The laser would burn off a proportion of the underside. The evaporating stream of material would then act as a thrust, nudging the particle from its circular orbit around the Earth into a more elliptical one before burning up. Lasers could nudge the debris off course and eventually into the upper atmosphere, where it would remain for millions of years. Project director Jonathan Campbell from the Advanced Concepts Group at Marshall Space Flight Center in Huntsville says, "The sizes we are aiming at are tough to detect and impossible to protect against."[38] It would take two years and between $50 million and $100 million to perform such a clearing operation.

NASA is planning to emit a gigantic 20-meter-wide laser beam from a massive earthbound device. Project Orion, as it is known, would be able to melt away half-inch chunks of rubbish 900 miles up in the atmosphere. Orion scientists also proposed a more expensive three-year strategy to clear all debris beneath an altitude of 1,500 kilometers, and it was completed in 2001.

The NASA laser scheme, planned jointly by NASA and USAF Space Command, would use radar to detect a suitable piece of orbital debris. The project would operate from USAF Space Command in a desert site. Scientists have called on state-of-the-art adaptive optics, with lenses and mirrors to deform and focus different parts of the beam, linked to atmospheric measuring devices.

Pieces of space debris that are larger than 10 centimeters apparently present no problem because they can be observed from the ground and the crew can take avoiding action. NASA's Orion project has been devised to obviate the 1 in 10 chance that the *ISS* will be holed between the years 2000 and 2010,[39] the odds increasing because of the shattering effect of tiny objects.

For example the *ISS* has a shield that will protect it from debris smaller than one centimeter across by exploding them into even tinier particles, which some say could actually increase the debris problem if the particles are micron-sized. Similarly, pieces in between 1 and 10 centimeters will generate a shower, several of which could penetrate the hull. "The result could be much worse, the difference between a single bullet and a shotgun blast," says Jonathan Campbell. Hence a "hoovering" laser for these small objects would be highly desirable, and orbits could be cleared within two years at a cost of just $200 million, he says.

Not everyone agrees with the "hoovering" projects. In 1995 the National Research Council could not see this in the offing. "Once we load these orbits with debris it is very difficult to get it out," says George Cleghorn, head of the panel that prepared a report for the Council and former chief engineer for TRW Space and Technology Group[40] in a definitive report that formed the basis for internationally agreed antidebris rules.

There are three awkward hurdles. First, the laser beam has to pass through a polluted, turbulent atmosphere that could deflect the laser beam. Second, the extremely narrow beam would itself tend to diverge as it journeyed towards its target, and its depleted energy would render it useless.[41] Could the laser also inadvertently damage satellites? Or be used to destroy undesirable satellites? Jonathan Campbell thinks that the laser beam would have to be fairly weak. Campbell says he was skeptical at the beginning. However he said in 1996, "Not only is it feasible in theory, but we already have equipment that would allow us to clear all the debris of that size range below an altitude of 800 kilometers." This 800 kilometer 'safe zone' would protect many valuable space assets, including many satellites fleets. At 500 kilometers, the manned space stations also fall into this zone.

Third, the problem with this type of missile approach, which depends on research from western military agencies, is whether international treaties would prevent the use of laser weapons in space.

Hence, Project Orion is only designed to test the ability of the laser to lock onto space junk. "There's certainly a track record of space debris being used as a cover story for missile defence projects," says John Pike, a well-respected policy analyst at the Federation of American Scientists in Washington. "Plenty of other countries have reasons to be suspicious," he adds.[42]

One way to test the laser project is to get an astronaut to push a piece of mock debris overboard from the shuttle. The debris would be so wired that it could be monitored from the ground. In June 2000 a series of minisatellites the size of footballs were launched from Baikonur in a test flight aimed at deorbiting space junk. Developed by Craig Underwood of Surrey Space Technology, a company owned by the University of Surrey, it was called "Snap" and broke new ground in space technology, to put Britain ahead of the world at the reasonable price of £100,000. He was the first to experience the perils of space junk since Underwood's team had built the ill-fated satellite that was hit by a bit of the Ariane rocket in 1996.

Weighing just 13 pounds, each minisatellite was fitted with an artificial brain and four video cameras each weighing no more than a quarter. They attached themselves to old satellites and other rubbish and propelled them into a graveyard orbit. Once homed in on a target, Snap used its butane-filled rocket motor to latch onto the debris. Craig Underwood said, "It costs almost as much to bring a satellite back to Earth as it does to put it up there. These satellites . . . will basically be on suicide missions."[43] The eventual plan is for swarms of Snaps to be launched to "spring clean" space on a regular basis.

SHIELDING AND CLADDING

Shielding is inevitably becoming a more important aspect of satellite design. This is the view of Mark Matney of Lockheed-Martin, who works at the Orbital Debris Program Office at the Johnson Space Center in Houston, Texas. Scientists at the Jet Propulsion Laboratory have high expertise in technologies needed for meteoroid protection, and are regularly testing shields.[44]

Exciting new opportunities for shield development arise through the rapid development of new materials such as sub-micron reinforced ceramics and other new fiber types. However, Aemonn Daly believes things have not been helped by intense rivalry between spacecraft manufacturers, who are using smaller, more vulnerable mass-marketed components unable to withstand the harsh space environment. He is concerned to redress this inadequacy by urging manufacturers to make onboard computers cleverer and able to recognize single-particle events, and to have better layout of equipment.[45]

Recent tests done by the ESA, Daimler-Chrysler Aerospace (DASA), and the Russian company NPO Lavotchkin have proved that an inflatable heat shield can protect satellites from the intense heat they face when reentering the atmosphere. This inflatable shield is cone-shaped, made of lightweight materials, and has been shown to withstand temperatures of 500°C.

Spacecraft are generally protected by what is known as a bumper shield, or a whipple shield, made of thin aluminum. The present generation of heat shields date back to the 1970s, when they were designed to protect spacecraft from overheating if they got too close to the sun. Double-carbon cladding was first used, consisting of layers of carbon "cloth" fibers baked in an oven with epoxy resin to form an extremely stiff but light shield. In other cases carbon fiber reinforced plastics (CFRP), multi-layer insulations and sandwich panels with CFRP facings, and aluminum honeycomb cores are used. Later shields were further strengthened at the manufacturing stage with methane gas, which deposits more carbon residue into the resin.[46]

The Russian components of the *ISS* fare slightly worse since they were designed before engineers started thinking seriously about the threat from debris. Extra shields were being deployed by NASA and the ESA to be used on their spacecraft, but only a few were fitted to the Russian components,[47] although scientists at the Russian space agency have developed on-board puncture detecting and repair kits. With the Proton launcher they propose to avoid the separation of propulsion units from the module.

In the meantime, for the other nations, the cost of "armor-plating" the *ISS* against man-made rubbish has added at least $5 billion to their bills.[48] The *ISS* will also contain advanced technology shields around habitable compartments, fuel lines, and control gyroscopes. As a speeding object approaches, it first encounters a sheet of aluminum about 2 millimeters thick, known as the Whipple bumper, which causes the projectile to shatter.

Space debris protection could also be made from Kevlar™ fabric, used by police and soldiers under fire and woven from flexible but unbreakable carbon filaments. Nextel™ foam cladding is another idea, designed to vaporize a particle into hot gases the instant it hits a spacecraft, thus dissipating its impact energy and blunting the particle before it can punch a hole.[49] The fragments are then slowed by more layers of Kevlar™, and finally the fragments bounce harmlessly off the spacecraft walls.

On the other hand, recent events have showed that Teflon™, the traditional cladding for spacecraft, is not as effective as was once thought. Although suspicious about Teflon™, NASA didn't expect to see a problem with the coating on the Hubble space telescope for 10 to 15 years. But cracks began to show on the thermal protection material in 1993, just three years after launch. In tests at NASA's Glenn Research Center in Ohio, Kim de Groh found that radiation makes Teflon™ brittle, leaving it less able to withstand the temperature extremes of space.[50]

Some satellite designers are starting to place sensitive instruments toward the back of the satellite to minimize impact risks. The side facing away from Earth is also shielded. Glass components such as the large area of coverglass used for the environmental protection of photovoltaic modules on the solar arrays can be strengthened. So far, the glass has to be thin (100 to 400 microns) for the economics of launch. Protective "bumpers" can be built, as they were for the Giotto mission, which penetrated into the head of Halley's comet.

The cupola of the *ISS*, developed by a European consortium in Turin headed by Aleania Aerospazio, has an outer unit that is specially strengthened and tested to withstand 8,600 pounds of pressure per square inch and is virtually impermeable to space debris.[51]

On behalf of the French SES, operators of the *Astra* satellite (more commonly known as the home of *BSkyB*), Pascale Reuland said that manufacturers were looking to soften the impact of micrometeorites between the layers to reduce the velocity of impacts. A polymer coating that conducts electricity has been developed by Canadian scientists to help satellites withstand solar storms.

Although in space solar cells have to withstand intense radiation, it is generally recognized that solar panels are impractical to shield. Scientists trying to make better cells have long been intrigued by a complex semiconductor called CIGS (copper indium gallium diselenide). This absorbs light more efficiently than silicon and is surprisingly stable even when exposed to intense radiation.

The secret of CIGS is its remarkable ability to repair itself, according to chemist David Cahen at the Weizmann Institute of Science in Tel Aviv. He and his researchers found that some chemical bonds can be easily broken, freeing copper atoms to wander through the crystals and repair radiation damage in the process. "The remarkable thing," says Cahen, "is that you can have a high-quality electronic material with these atoms running around."[52] The natural tendency of the copper atoms to distribute themselves evenly means they spread into damaged spots in the semiconductor crystals. Robert Tomlinson, a materials scientist at the University of Salford in Manchester, says these self-repairing CIGS-made solar cells could "provide the stable, radiation-hard power supply for long-term space missions." At least one company, Siemens, markets CIGS solar cells, which generate electricity more efficiently than other thin-film solar products.

But at such hypervelocity speeds, the object, should it strike head-on, could cause the module's skin to melt. According to Steve Hall of NASA's Marshall Space Flight Center, providing the hole that was blown in the *ISS*'s shell in the advent of a debris strike was no bigger than 10 centimeters across, the astronauts could survive.[53] Eric Christiansen, chief analyst at the Hypervelocity Impact Test Facility at the Johnson Space Center, says that most puncture wounds from a piece of debris the size of a coin would allow the atmosphere to leak out only slowly, giving the astronauts time to fix the hole Band-Aid fashion.[54] If this didn't happen, a newly developed puncture repair kit would allow the astronauts to patch up the hole during a space walk.

NASA's puncture patch is made of a clear, polycarbonate disc that can be bolted to the module's skin while on a space walk. An epoxy resin is then squirted into the disc using a gun that mixes resin with a hardener. This will give six months grace. Says Hall, "Permanent repair techniques are highly dependent upon the specific pattern of damage sustained, and they are being studied separately." He adds: "They may resemble the type of repairs made on aircraft fuselages." Walter Flury of the ESA agrees that such an exterior repair mechanism is vital because not all walls are accessible from the inside.

The ESA, which plans to launch the Columbus Laboratory section of the Space Station in 2003, is also studying puncture repair techniques.[55] NASA and NASDA (of Japan) have in the past used bolt catchers and special tethers to limit the release of debris, and have jettisoned fuel on dead spacecraft to prevent explosions in space.[56] Other mechanisms are still at the blueprint stage, including giant foam balls. A particle penetrating the foam would lose energy and fall back to Earth sooner.

In the meantime, NASA was waiving its safety requirements to spur things along, since the crew module had failed NASA's shield tests for surviving collisions with orbiting space junk.[57] Indeed, the shields were not ready until 2002, as hinted at by a spokesman for the General Accounting Office when addressing a Senate panel overseeing NASA's activities, possibly reflecting Russia's method of cutting corners. *Mir*, however, was in orbit for 15 years with debris shields, and no serious space debris punctures or accidents occurred, says Joe Rothenberg, NASA's associate administrator for space flights.[58]

No proof could be better than surviving actual conditions of space. At present, and for the time being, existing cladding and shielding devices seemed to be working quite well.

CHAPTER 11

Defend the Earth

How do the space authorities plan to counter the impact of much larger space objects, such as the meteoroids? Clearly no amount of shielding or cladding will save a spacecraft—it will be destroyed in an instant in such a collision.

The most frequent suggestion about countering the asteroid threat is to try to destroy it in space before it reaches Earth's surface, as depicted in the movies *Armageddon* and *Deep Impact*. Alternatively, the object could be deflected from its orbit, which could also be done with nuclear weapons fired from a rocket that reached speeds of at least 40,000 mph. The weapon would not need to strike the asteroid head-on; it would simply need to fly level with it and nudge it—by exploding a nuclear warhead—into a different trajectory. Indeed, the Rutherford Appleton Laboratory (RAL), along with other space institutions, has given much thought to this "deflection" technique.[1]

However, difficult calculations would have to be made about the ecliptic plane of the Near Earth Object (NEO) that is threatening Earth for a launch vehicle to travel alongside it. Lengthy verification and planning would be needed for any type of militaristic space venture. The size of the cosmic object essentially depends on the larger decision of whether to vaporize it or to deflect it. Professor Robin L.

Kirk of the College of Aeronautics, Cranfield, Bedfordshire, is skeptical about getting nuclear missiles to rendezvous with a high-speed incoming missile, which travels much faster than ICBMs, at up to 45,000 mph.[2] The impacter's orbit would need to be determined way in advance, and the rocket launched with incredible accuracy before the missile gets too close to Earth. The long time gap—up to eight years—before the interception could take place is a decided drawback. Professor Kirk points to the failure of the Patriot missiles during the Gulf War. The break-up of an asteroid has also to be avoided to prevent the scattering of Earth-bound debris. Again, a relatively tiny object, say less than a mile across, has to be detected against a background of stars. "A small navigation error on the interceptor would result in mission failure, requiring more missiles to be launched," he wrote.[3] Further, the estimated trajectory will be known only days before the event.

Nuclear explosives may be needed for objects wider than 100 meters that are spotted late in the day and intercepted at a distance no closer than about 150 million kilometers, since the deflecting or destructive energy needed is about equivalent to that yielded by a one-megaton nuclear explosion. Closer to Earth than 150 million kilometers and the megatonnage goes up phenomenally, says Gregory Canavan, a senior scientist at Los Alamos National Laboratory in New Mexico.[4]

Jay Melosh, of the Planetary Sciences Department of Arizona University, said that nuclear warheads 100 times the size of those available today would be needed: "The threat of them being used on Earth could be greater than that of an asteroid."[5] For a small object known about years in advance conventional explosives could be used simply to deflect it. However, in terms of the long-period comets (appearing relatively suddenly to astronomers) literally nothing could be done about them. Focusing solar beams at a missile, for example, only works with a relatively long lead time.

The astronomer Gerrit Verschuur stresses that we must know in advance what the NEO or NEA is made of,[6] and how many of them are out there. Scientific experiments to this end might never stop, because the population of NEAs "is constantly evolving." You would have to decide in advance whether the space missile was solid, loose, or made of iron or rock, says physicist Edward Tagliaferri, a U.S. space program consultant.[7] Indeed, identifying all the potential threats from outer space will need a 10-year $50 million program, says David Morrison of the NASA Ames Research Center in California.[8] Still, that's less than the budgets of the two asteroid disaster movies made and shown in 1998.

Two NASA workshops, in 1993, recommended detecting techniques for identifying incoming objects, and similar methods regarding how objects could be intercepted. Photographs alone are not enough, says Morrison.[9] NASA scientists proposed building six 2.5 meter telescopes, three located in the north and three in the southern hemisphere. Each would be equipped with advanced versions of the Charge Coupled Device (CCDS), a kind of electronic camera already being used on some missions. The CCDS system records electronic images of space objects and then feeds them into computers. The CCDSs could, it is suggested, find most of the 300,000 ECAs of about 100 meters or less within a decade, and virtually all of the others within 25 years.

Let us assume that the lead time of the impending missile is relatively long, which generally applies to NEOs. A "precursor mission" similar to Clementine should be sent first—an instrument spacecraft to fly up and literally search for them. A Titan rocket will launch a spacecraft into orbit ten miles above Earth with a payload of three probes, each one meter long. For a year the satellite will track asteroids and on command from Earth will release the probes. Optical navigation sensors on the craft will lock on to them and strike them at speeds of up to 11 mps.[10]

DESTROYING BOLIDES—TOO DANGEROUS?

Nevertheless, even in theory the task of destroying incoming bolides is daunting. Detonating a device at the wrong moment or position would fragment the object, creating smaller objects, or wrongly shift its orbit toward a populated area. Gerrit Verschuur, at a UN Conference on NEOs in April 1995, asked what we could expect if the nuclear standoff explosion failed to deflect the asteroid, and instead merely loaded the object with radioactivity instead of "harmless" rock. He says it might be better to live with the consequences of fragments arriving after a stand-off nuclear explosion rather than the giant asteroid hitting earth.[11]

There could also be considerable political and social opposition to the launching of nuclear missiles, and complicated issues concerned with nonproliferation and ABM treaties. In the early 1990s a draft copy of a report by a NASA committee made startling reading. It mentioned setting up laser guns on both the Earth and the moon, as well as establishing an "armada" of hundreds of Earth-orbiting rockets, each carrying a 100-megaton warhead or even antimatter guns.[12] It suggested we start right now with "target practice" against asteroids, using nuclear weapons. The Clementine II project, sponsored by the USAF, had a dual military function, as many critics of these types of space programs often allege, of testing new missile guidance systems by destroying asteroids. The USAF had set aside $120 million (in the mid-1990s) for the project, hoping to keep the Air Force in the Star Wars game while enabling its scientists to study the latest tracking technology.

Many people looked on with alarm when the esteemed physicist Edward Teller made similar suggestions.[13] Other scientists, such as the late Carl Sagan, in view of the notion that perhaps the small meteoroids of just 50 meters wide or less would provide the defense industry with a battery of cosmic targets to practice on, have warned of the dangers of disguising the deployment of nuclear and ballistic technologies in this way. Duncan Steel has also pointed to the danger of Spaceguard conferences with participants "angling for new big-budget programs and the power that goes with them." There is always the danger of the "sensible scientist versus the loopy weaponeers" debate clouding the issue.[14]

As it happened the NASA committee's "nuclear option" was quietly dropped from a brief preliminary report submitted to Congress in March 1991, and one committee member, Clark Chapman, of the Planetary Science Institute in Tucson, Arizona, insisted his name be removed from the report because of the implications involved in the use of military technologies. But in October 1997 President Bill Clinton vetoed the *Clementine II* mission, intended for smashing into *Toutatis*, scheduled for early 1999.

However, it is interesting to note the comments of Greg Canavan, who points to the alarming global publicity given to the comet Shoemaker-Levy as it broke into the equivalent of million-megaton-bomb pieces, which plunged, one after the other, into the atmosphere of Jupiter in 1994 and were clearly visible through Earth telescopes. "Nothing clears the mind as the sight of the gallows," quipped Canavan.[15]

Gerrit L. Verschuur, in his book *Impact*, puts it another way. "It would be ironic if in the final analysis ill-informed opponents of a plan to deflect a rogue asteroid were to obtain a restraining order on the launch, in which case one of the last headlines ever to be published in the world might read 'Court Prevents Launch of Asteroid Bomb.' "[16]

Aware of the strong antinuclear sentiment among members of the U.S. Congress, scientists at the Lawrence-Livermore National Lab in California instead came up with a scheme borrowed from the Star Wars "brilliant pebbles" strategy coupled with conventional mining techniques. The nose-cone of a warhead, instead of being nuclear-tipped, could be packed with an array of interconnected tungsten balls and springs. At the appropriate moment the balls would be jettisoned by the springs and would bury themselves deep into the asteroid to convert kinetic energy into thermal energy. This, surprisingly, would generate about 100 times more destructive power than high explosives. "You'd get an array of explosions going off and the shocks between them would propagate out and fracture all the rock between the points of explosion," explains Canavan.

NEUTRON BOMBS AND PLASMA BEAMS

Others at the Los Alamos and Lawrence Livermore Labs have toyed with other ideas.[17] To avoid fragmenting a stray asteroid into large chunks, a neutron bomb could be used. Neutron bombs deliver most of their energy in form of speeding neutrons rather than an explosive blast. The neutron warhead could be detonated when the missile approached to about a distance equal to the radius of the asteroid, says Greg Canavan. The neutrons would heat and vaporize the material, and blow out one side of the asteroid.

Still other ideas have come from the Deep Impact Mission team, which includes NASA's Jet Propulsion Laboratory in California and Ball Aerospace Technologies Corporation in Colorado. The plan is for a robotic spacecraft that will blow up part of a comet with a huge copper bullet in the year 2004. This would be a purely scientific mission to attempt to examine the nature of comets, but prove that cometary defense technologies are feasible. When the "impacter" projectile, which will weigh 771 pounds, collides at 22,300 mph with the comet's icy core, it will blast a hole seven stories deep and as wide as a football pitch, releasing clouds of debris into space.[18]

In another project, sometimes known as "the plasma lab in the sky," HAARP (standing for High-frequency Active Auroral Research Project) can beam radio waves into the ionosphere to produce energy densities roughly equal to what the Earth naturally produces in the

Extremely Low Frequencies (ELF) range, according to its program managers, and these could also be focused on incoming asteroids. It is both a weapons, space research, and weather modification technology. Other reasons for HAARP's existence include investigating the way the sun interacts with the atmosphere. Official documents speak of "a major Arctic facility for upper atmospheric and solar-terrestrial research."[19] Like the nuclear-warhead deflector, HAARP has dual purposes. Thermonuclear detonations in the upper atmosphere operate on similar power levels and produce similar effects. In fact, HAARP can create the same kind of EM pulses as do nuclear explosions.[20]

Because the HAARP project appears to be a powerful Department of Defense facility and is paid for out of the Office of Naval Research budget, and because it is jointly managed by USAF Phillips Geophysics Lab based at Hanscom AFB, in Massachusetts, some have suggested it is a prototype for Star Wars–type weapons.[21]

A series of four high mountains form part of the Wrangell-St. Elias National Park. Against one of the mountains, Mt. Sanford and the Cooper River Valley, there are four dozen 72-foot-tall metal towers, crisscrossed with guy wires and metal meshes. The site is on the Alaska Route 4, about 260 miles northeast of Anchorage. The total acreage makes its size comparable to six Yellowstone Parks, with a massive grid completed in 2002. It has 30 transmitter shelters, each housing 12 diesel-powered transmitters each capable of generating 10,000 watts of radio frequency power (RF), totalling 3.6 million watts in all. HAARP can focus these transmissions into a single point in the sky, achieving in excess of 3.6 billion watts.[22]

Other antibolide proposals include tying a rocket or an EM gun to an asteroid, to cause "outgassing," whereby cracks created in the surface of the rock by a small explosion would cause volatile material to escape and act as a deflecting jet.

Yet another suggestion, put forward by Jay Melosh and by Ivan Nemchinov at the Institute of Geosphere Dynamics in Moscow, describes a vision of a giant solar sail, between one and ten kilometers in diameter, modified to focus intense beams of sunlight, with a secondary mirror aiming the beam onto the asteroid. Hot gases generated would evaporate the loose silicate rocks and ice, melting on the asteroid's surface at up to 2,000°C, which would then deflect its trajectory in a steady push.[23] The whole ensemble would be launched from a shuttle or rocket on a path that kept pace with the rogue asteroid. A half-kilometer-wide mirror could deflect an asteroid of

2.2 kilometers in a year, or 10 kilometers in a decade, equivalent to about 1,000 megatons of nuclear blast.[24] NASA's deflection programs are based on the understanding that asteroids are not solid masses of rock but lumps of materials easily pried apart by an impact or a bomb exploding close to the object. The Rutherford Appleton Laboratory in Oxford, working with the ESA, has already embarked on the Rosetta mission, designed for a rocket to move in formation with any incoming natural missile, and other missions have actually landed on the surface of comets and asteroids (e.g., as with the comet Wirtanen and the asteroid Eros).

NASA's spacecraft *Clementine II*, budgeted by the U.S. Defense Department in coordination with NASA, has had some success in rendezvousing with asteroids. It flew an instrument package past the approaching asteroid Geographos to test the kind of sensors and navigational devices needed. There were also plans to send *Clementine II* on a dramatic mission to the asteroid Toutatis—two miles across—to first take pictures of it before slamming into it at 45,000 mph to give some hint as to its composition and strength of cohesion.[25]

Scientists at Johns Hopkins Applied Physics Laboratory have plans for a "soldier" or "killer sentinel" deflection spacecraft positioned at strategic spots in near space, enabling interception and destruction within a year of spotting an intruder of about 300 meters wide, although present evidence suggests that if it were designed actually to land on the asteroid, ten years notice would be needed.[26]

Tom Gehrels, a University of Arizona astronomer who heads one of three U.S. teams searching for asteroids, reckons 100 million asteroids larger than 20 meters in diameter are on earth-crossing orbits, so that they would be difficult to detect.[27] Most "lead times"—the length of time astronomers would be aware of a missile heading toward Earth—would only be weeks in the case of long-period comets. Long-period comets, returning, for example, after 200 years, appear without warning. Then only a few months of notice might be given. Brian Marsden of the Harvard Smithsonian Center for Astrophysics says proper searches should give plenty of warning of all threats except from long-period comets from beyond the solar system. But these only amount to about 4 percent of the threat. If the comet is on a 217,000 km/h collision course, a quick nuclear bang would be needed.[28]

Many believe that any deflector should only be launched from an orbital platform to reduce the need for enormous launch energies that would require a rocket to escape Earth's gravity. Ion drives could be used instead of chemical rocket fuels, which are especially useful once outside Earth's gravitational field. NASA's *Deep Space 1* was launched in 1998 using an ion drive, which demanded 2,500 watts of power to generate its maximum thrust.[29]

THE SPACEGUARD PROJECT

However, in spite of reservations in some quarters about the use of "Star Wars"–type missiles to destroy incoming meteoroids, this option has not been ruled out by influential groups of astronomers. After much campaigning by the influential Spaceguard U.K. pressure group, the British government took seriously the threat to Britain and indeed the world of the consequences of a large meteoroid impact—and the need for them to be deflected by military-type systems.

In September 2000 it published its Task Force report and recommended, prior to considering the military option, the building of a new 3-meter-class survey telescope to search for Near Earth Objects (NEOs). It drew attention to other surveillance regimes actually in operation that could be utilized for more focussed NEO observations, especially the ESA's Gaia mission, and NASA's Space Infrared Telescope Facility (SIRTF). It also recommended the global coordination of sky searches with inexpensive microsatellites, and the coordination—with a major British input—of other NEO research and observation studies and forums, to formulate Spaceguard strategies. One of its recommendations does indeed suggest that the United Kingdom and other governments "set in hand studies to look into the practical possibilities of mitigating the results of impacts and deflecting incoming objects."

This latter aspect was an appeal for spending as much as $100 million on a defense system against space objects, similar to a missile defense system, with an early warning network coupled with some means to stop an incoming threat. The astronomer Duncan Steel says that such a leading role for Britain (presumably referring only to the observational aspect) would cost U.K. taxpayers some £10 million annually, but would be well worth the trade-off in view of the massive economic damage that would occur to the United Kingdom should the meteorite strike. He mentions other monitoring assets already in operation that could be linked up with the British scheme, such as the Visible and Infrared Survey Telescope for Astronomy (VISTA) project, which comes on line in 2004, the telescope on La Palma, and another in Hawaii.

Lord Sainsbury, the British science minister, said: "This is not science fiction. The risk is extremely remote, but it is real. We put a lot of money into astronomy. It's sensible to put just a little bit into making certain we know if there is any danger of an object hitting our very fragile planet."[30]

Louis Friedman, of the Planetary Society, pointed out that further vigilance is required because our present observations are based on only 10 percent of possible NEOs. "Whether there is a threat or not, or what the nature of it might be, depends on details . . . we just don't have."[31] He added that the task force report adopted the right cautionary tone without resorting to "hysterical doomsday language," although he is surprised that Britain should be pushing for its own space-based observatories since Britain's own space science exploration programs are rather meager compared with America's or Russia's, and Russia seems to have been left out of the picture in regard to the international Spaceguard effort.

But it will only be a matter of time before the excellent astronomical facilities of the Russians will also come on board the Spaceguard Project. In any event the global nature of meteoroid search is becoming more notable and is beginning to rival climatic change as a universal cause for concern. Astronomer Brian Marsden of the International Astronomical Union (IAU) refers to the long-standing search for bolides and asteroids dating back over 30 years in the United States, such as the Lincoln Near Earth Asteroid Research (LINEAR) system, Spacewatch, Near-Earth Asteroid Tracking (NEAT), and Catalina Sky Survey (CSS) among other acronymic monitoring systems using telescopic, photographic, and other high-tech observational procedures.

The Spaceguard Foundation, formed as a result of decisions spurred by the Council of Europe's Resolution 1080 of 1996, and of other decisions taken by the IAU, has appointed Britain's new Spaceguard Centre in Wales, set up in 2001, as the International Spaceguard Information Bureau. Led by Jonathan Tate, it has become the leading public outreach organization, consisting of astronomers, analysts, and risk experts, as well as government officials, dealing with NEOs in the world's skies. In an interesting aside, Andrea Carusi, president of the Spaceguard Foundation, justified continued pressure on governments by pointing to the difference between "normal" scientific activity and a dreadful catastrophe. "We don't need to be alarmed about the cosmic threat," he said, "but we *do* need to be concerned."[32] This resulted in the Council of Europe's "famous" Resolution 1080 of 1996, which adopted similar resolutions to those of Spaceguard.

Notes

CHAPTER 1: THE *MIR* FIREBALL

1. *Sunday Times*, 18 March 2001, p. 1.24.
2. *International Herald Tribune*, 15 March 2001, p. 9.
3. *Sunday Times*, 11 March 2001.
4. Ibid. p. 1.20.
5. *International Herald Tribune*, 23 March 2001, p. 9.
6. *The Times*, 29 August 1998, p. 24.
7. Ibid.
8. *The Times*, 27 December 2000, p. 5.
9. *New Scientist*, 25 November 2000, p. 6.
10. *International Herald Tribune*, 24 March 2001, p. 5.
11. *Sunday Times*, 18 March 2001, p. 1.24.
12. *The Times*, 24 March 2001, p. 16.

CHAPTER 2: HAZARDOUS SPACE MISSIONS

1. Christopher Lee, *War in Space*. Sphere Books, 1987, p. 30.
2. *Times Interface* (suppl.), 1 October 1997, p. 13.
3. James Davies, *Space Exploration*. Chambers, 1992, p. 236.
4. Ibid., p. 242.
5. *Quest* (UK), August–September 1997, p. 58.

6. *New Scientist*, 17 June 2000, p. 17.
7. *New Scientist*, 9 May 1998, p. 4.
8. *Unopened Files* (UK), January–February 2000, p. 44.
9. *The Times* (UK), 30 October 2001, p. 17.
10. *Focus* (UK), February 2000, p. 75.
11. *Sunday Telegraph* (UK), 9 August 1998, p. 23.
12. *New Scientist*, 4 May 1999, p. 44.
13. *The Times* (UK), 17 September 1997, p. 13.
14. *The Observer* (UK), 21 June 1999, p. 2.
15. *New Scientist*, 22 January 2000, p. 20.
16. *Sunday Times* (UK), 4 February 2001, p. 1.2.
17. *International Herald Tribune*, 24 March 2001, p. 5.
18. *New Scientist*, 4 May 1999, p. 41.
19. *Daily Telegraph* (UK), 14 September 1997, p. 26.
20. *International Herald Tribune*, 14 December 1997, p. 26.
21. *Frontiers* (UK), February 1998, p. 37.
22. *New Scientist*, 5 September 1998, p. 37.
23. *New Scientist*, 12 September 1998, p. 5.
24. *The Times* (UK), 11 September 1998, p. 18.
25. *Frontiers* (UK), November 1998, p. 39.
26. *New Scientist*, 18 September 1999, p. 5.
27. *The Times* (UK), 13 July 2000, p. 18.
28. *New Scientist*, 16 May 1998, p. 12.
29. *International Herald Tribune*, 4 December 1999, p. 5.
30. *New Scientist*, 12 July 1997, p. 12.
31. *New Scientist*, 20 November 1999, p. 5.
32. *New Scientist*, 17 June 2000, p. 16.
33. *The Times* (UK), 11 February 2000, p. 22.
34. *New Scientist*, 19 February 2000, p. 4.
35. *New Scientist*, 1 September 2001, p. 11.
36. *International Herald Tribune*, 4 December 1999, p. 5.
37. *International Herald Tribune*, 17 January 2000, p. 3.

CHAPTER 3: MAN-MADE MISSILES FROM SPACE

1. James Davies, *Space Exploration*. Chambers, 1992, p. 111.
2. *Quest* (UK), October 1998, p. 60.
3. *Something in the Air*. Robert Hale, 1998, p. 176.
4. *Geophysical Review Letters*, vol. 28, p. 959.
5. *The Times* (UK), 17 May 2000, p. 31.
6. See *UFO Magazine* (UK), January–February 2002.
7. *Sunday Times* (UK), 26 March 2000, p. 1.27.
8. *New Scientist*, 4 March 2000, p. 5.
9. *Frontiers* (UK), January 1999, p. 101.

10. *Focus* (UK), February 2000, p. 76.
11. *International Herald Tribune*, 15 March 2001, p. 9.
12. *New Scientist*, 22 January 2000, p. 5.
13. *New Scientist, Business in Space* (suppl.), 24 May 1997, p. 1.
14. *Focus*, September 2000, p. 8.
15. *Sunday Times* (UK), 4 February 1996.
16. *New Scientist*, 14 November 1998, p. 41.
17. *New Scientist*, 24 April 1999, p. 19.
18. Proceedings of the Third European Conference on Space Debris, SP-473, p. 527.
19. Ibid., p. 541.
20. *Frontiers*, October 1998, p. 20.
21. Third Conference, p. 356.
22. *New Scientist*, 17 June 2000, p. 17.
23. *New Scientist*, 24 October 1998, p. 40.
24. *New Scientist*, 9 May 1998, p. 4.
25. *New Scientist*, 8 November 1997, p. 14.
26. *New Scientist*, 31 October 1998, p. 24.
27. *International Herald Tribune*, 4 September 2001, p. 11.
28. Third Conference, p. 477.
29. Ibid., p. 464.
30. *International Herald Tribune*, 12 April 2000, p. 18.
31. *New Scientist*, 25 July 1998, p. 42.
32. *New Scientist*, 25 March 2000, p. 9.
33. *International Herald Tribune*, 12 April 2000, p. 18.
34. Third Conference, p. 496.
35. Ibid., p. 496.

CHAPTER 4: TOXIC RAIN FROM SPACE

1. Matt Irvine, *Tele-Satellites*. Gloucester Press, 1989, p. 17.
2. *New Scientist*, 22 January 2000, p. 18.
3. *New Scientist*, 6 March 1999, p. 40.
4. Ibid., p. 41.
5. *Scientific American*, August 1998, p. 44.
6. *Nexus* (Australia), August–September 1998, p. 7.
7. Jack Challoner, *Space*. Channel-4 Books, 2000, p. 115.
8. ESA Third Conference, SP-473, p. 489.
9. *New Scientist*, 18 September 1999, p. 5.
10. *New Scientist*, 6 November 1999, p. 25.
11. *Frontiers*, September 1999, p. 38.
12. *New Scientist*, 11 October 1997, p. 18.
13. Ibid.
14. *The Times* (UK), 8 September 2000, p. 20.

15. *New Scientist*, 6 March 1999, p. 42.
16. Ibid.
17. *Scientific American*, September 1999, p. 11.
18. *Sunday Times* (UK), 14 September 1997, p. 1.26.
19. *Ufo Reality* (UK), August–September 1997, p. 17.
20. Challoner, *Space*, p. 117.
21. *Unopened Files* (UK), December 1999, p. 67.
22. *Alien Encounters* (UK), issue 25, 1997, p. 39.
23. *The Times* (UK), 12 January 2000, p. 16.
24. *Ufo* (UK), July–August 1997.
25. *Spaceflight*, March 1995, vol. 37.
26. *Alien Encounter* (UK), issue 25, 1997, p. 39.
27. ESA Third Conference, p. 863.
28. Ibid., p. 869.
29. Ibid., p. 870.

CHAPTER 5: THE MOONDUST PROJECT

1. Donald Keyhoe, *Aliens from Space*. Panther Books, 1975, p. 86.
2. Nicholas Redfern, *Cosmic Crashes*. Simon & Schuster, 1999, p. 283.
3. *Quest* (UK), document QP 217, 1998.
4. Redfern, *Cosmic Crashes*, p. 31.
5. Jenny Randles, *Something in the Air*. Robert Hale, 1998, p. 15.
6. *Quest* (UK), document QP 102, p. 9.
7. Ibid.
8. *Ufo* (UK), September–October 1998, p. 72.
9. Redfern, *Cosmic Crashes*. p. 286.
10. *Quest* (UK), document QP 102, pp. 1–5.
11. Ibid., p. 27.
12. Ibid., pp. 29–30.
13. Ibid., pp. 22–23.
14. Ibid., p. 15.
15. Ibid., pp. 19–20.
16. Ibid., p. 17.
17. Ibid., p. 43.
18. Ibid., p. 25.
19. Jack Challoner, *Space*. Channel-4 Books, 2000, pp. 113–14.
20. ESA Third Conference, SP-473, p. 501.
21. Ibid.
22. Ibid., p. 505.
23. Challoner, *Space*. p. 112.
24. *Times Higher Education* (suppl.), 20 November 2000, p. 24.
25. Timothy Good, *Above Top Secret*. Acadia, 1989, p. 218.
26. Nicholas Redfern, *Covert Agenda*. Simon & Schuster, 1997, p. 161.

27. Ibid., p. 173.
28. Jenny Randles, *Something in The Air*, Robert Hale, 1998, p. 164.
29. Ibid., p. 106.
30. Ibid., p. 107.
31. Ibid., p. 166.
32. *Ufo* (UK), June 2001, p. 40.
33. Randles, *Something in the Air*, p. 194.
34. *Northern Ufo News*, issue 62, 1993.
35. *Ufo* (UK), May–June 1998, p. 71.
36. Ibid., June 2001, p. 38.

CHAPTER 6: THE JUNKYARD IN SPACE

1. *Scientific American*, August 1998, p. 46.
2. Ibid., p. 42.
3. Ibid., p. 44.
4. Ibid., p. 46.
5. Proceedings of the Third European Conference on Space Debris, SP-473, p. 754.
6. *Times Higher Education* (UK), 8 January 1999, p. 12.
7. *New Scientist*, 24 July 1999, p. 17.
8. Proceedings of the Second Conference on Space Debris, Darmstadt, SP-393, 1997, p. 368.
9. *New Scientist*, 27 May 2000, p. 38.
10. *The Times* (UK), 23 January 1999, p. 15.
11. Ibid.
12. *Ufo* (UK), July–August, 1997.
13. Arthur C. Clarke, *By Space Possessed*. Gollancz, 1993, p. 108.
14. *Times Higher Education* (UK), 16 July 1999, p. 31.
15. Ibid.
16. ESA Third conference, p. 765.
17. Ibid., p. 768.
18. *New Scientist*, 18 August 2001, p. 18.
19. *Focus* (UK), September 2000, p. 41.
20. *Times Higher Education* (UK), 31 March 2000, p. 3.
21. *New Scientist*, 24 July 1999, p. 14.
22. Ibid., p. 3.
23. Proceedings of the Second Conference, p. 78.
24. Ibid., p. 581.
25. Ibid., p. 38.
26. James Davies, *Space Exploration*. Chambers, 1992, p. 191.
27. *Scientific American*, August 1998, p. 44.
28. Ibid.
29. *New Scientist*, 6 March 1999, p. 44.

30. Proceedings of the Second Conference, p. 93.
31. *New Scientist*, 17 May 1997, p. 49.
32. Proceedings of the Third Conference, p. 355.
33. Ibid., p. 207.
34. Ibid., p. 548.
35. Proceedings of the Second Conference, p. 597.
36. *New Scientist*, 24 August 1996, p. 5.
37. See *Frontiers* (UK), Spring 1998.
38. *New Scientist*, 7 March 1998, p. 23.
39. Proceedings of the Third Conference, p. 262.
40. Ibid., p. 757.
41. Jack Challoner, *Space*, Channel-4 Books, 2000, p. 45.
42. *New Scientist*, 24 June 1995.
43. Proceedings of the Third Conference, p. 272.
44. *Sunday Times* (UK), 20 October 1996.
45. *New Scientist*, 24 June 1995.
46. Davies, *Space Exploration*, p. 111.
47. *Quest* (UK), October 1998, p. 60.
48. *New Scientist*, 27 May 2000, p. 33.
49. *Astronomy* (UK), December 2000, p. 56.
50. Proceedings of the Third Conference, p. 329.
51. *The Times* (UK), 25 March 1999, p. 20.
52. See *New Scientist*, 24 June 1995.
53. *Scientific American*, August 1998, p. 45.
54. *New Scientist*, 23 July 1994, p. 23.
55. Ibid., p. 94.
56. *Scientific American*, August 1998, p. 43.
57. Challoner, *Space*, p. 100.
58. *Scientific American*, August 1998, p. 43.
59. Proceedings of the Third Conference, p. 192.
60. *Frontiers* (UK), February 1998, p. 27.
61. *Guardian* (suppl.) (UK), 19 January 1999, p. 9.
62. *New Scientist*, 23 May 1997, p. 11.
63. Ibid., 4 September 1999, p. 44.
64. Ibid., 1 April 2000, p. 11.
65. Ibid., 6 November 1999, p. 5.
66. Ibid., 14 November 1998, p. 40.
67. *Sunday Times* (UK), 9 May 1999, p. 5.11.
68. Proceedings of the Third Conference, p. 197.
69. *Astronomy* (UK), December 2000, p. 58.
70. *Guardian* (UK), 19 January 1999, p. 9.
71. *New Scientist*, 22 September 2000, p. 5.
72. Proceedings of the Third Conference, p. 36.

CHAPTER 7: SPACE DEBRIS. THE DATA PROBLEM

1. *Astronomy* (UK), December 2000, p. 61.
2. *New Scientist*, 20 October 2001, p. 59.
3. *New Scientist*, 11 August 2001, p. 5.
4. Ibid.
5. Proceedings of the Second Conference on Space Debris, Darmstradt, SP-393, 1997, p. 367.
6. Proceedings of the Third European Conference on Space Debris, SP-473, p. 13.
7. Proceedings of the Second Conference, p. 39.
8. *Times Higher Education* (suppl.) (UK), 16 July 1999, p. 31.
9. *International Herald Tribune*, 20 June 2001, p. 17.
10. Proceedings of the Second Conference, p. 97.
11. Jack Challoner, *Space*. Channel-4 Books, 2000, p. 97.
12. *Sunday Times* (UK), 20 November 1996.
13. Proceedings of the Third Conference, p. 842.
14. James Davies, *Space Exploration*. Chambers, 1992.
15. *Quest* (UK), January 1998, p. 53.
16. Proceedings of the Second Conference, p. xv.
17. Proceedings of the Third Conference, p. 641.
18. Ibid., p. 36.
19. *International Herald Tribune*, 12 April 2000, p. 18.
20. Proceedings of the Second Conference, p. 601.
21. Proceedings of the Third Conference, p. 455.
22. *Scientific American*, August 1998, p. 44.
23. Proceedings of the Second Conference, p. 217.
24. *Unopened Files* (UK), December 1999, p. 59.
25. Proceedings of the Second Conference, p. 601.
26. Ibid., p. 93.
27. Challoner, *Space*, p. 104.
28. *International Herald Tribune*, 10 June 2001, p. 17.
29. *Scientific American*, August 1998, p. 46.
30. *Fortean Times* (UK), October 1997, p. 154.
31. Proceedings of the Second Conference, p. 217.
32. Proceedings of the Third Conference, p. 45.
33. Ibid., p. 45.
34. *New Scientist*, 4 July 1998, p. 32.

CHAPTER 8: THE SPACE INVADERS

1. *New Scientist*, 7 June 1997, p. 10.
2. Ibid., 11 November 2000, p. 17.

3. *The Guardian* (UK), 4 September 1998, p. 2.

4. *Evening Standard* (UK), 4 April 2001, p. 20.

5. *New Scientist*, 30 May 1998, p. 3.

6. Austen Atkinson, *Impact Earth*. Virgin Publishing, 1999, p. 126.

7. Proceedings of the Third European Conference on Space Debris, SP-473, p. 163.

8. Ibid., p. 164.

9. *Alien Encounter* (UK), December 1997, p. 36.

10. Atkinson, *Impact Earth*. p. xiv.

11. Proceedings of the Third Conference, p. 163.

12. Proceedings of the Second Conference, p. xxxvii.

13. *International Herald Tribune*, 20 June 2001, p. 17.

14. See *Ufo* (UK), Jan–Feb 2002.

15. *The Times*, 8 January 2002, p. 11.

16. *Spaceguard Impact* (UK), bulletin 9, p. 13.

17. *New Scientist*, 30 May 1998, p. 3.

18. "The Leonid 1998 Shower: Information for Spacecraft Operators," European Space Agency, Mission Analysis section, 1998.

19. *New Scientist*, 13 May 2000, p. 15.

20. "BBC Online," 29 November 1999, *Spaceguard Impact*, bulletin no. 9, January 2000.

21. Atkinson, *Impact Earth*, p. xxxxv.

22. Ibid., p. xxxiv.

23. "BBC Online," *Spaceguard Impact*, bulletin no. 9, 29 November 1999.

24. *New Scientist*, 14 November 1998, p. 43.

25. *Daily Mail* (UK), 29 October 2001, p. 34.

26. *Nature* (UK), vol. 405, May 2000, p. 321.

27. *New Scientist*, 20 May 2000, p. 38.

28. "BBC Online," *Spaceguard Impact*, bulletin no. 9, November 1999.

29. *Discover* (UK), November 1994, p. 31.

30. *The Guardian* (UK), 6 March 1997.

31. *New Scientist*, 11 November 2000, p. 17.

32. *Alien Encounter* (UK), issue 25, 1997, p. 8.

33. Ibid., issue 24, p. 8.

34. *Ufo* (UK), March–April 1999, p. 34.

35. *Alien Encounter* (UK), December 1997; *Sunday Times* (UK), 28 December 1997.

36. Atkinson, *Impact Earth*, p. 102.

37. *The Times* (UK), 17 December 1997, p. 15.

38. *Daily Mail* (UK), 17 December 1997.

39. *Sunday Telegraph* (UK), 28 December 1997, p. 11.

40. *Focus* (UK), June 1997.

41. *Ufo Reality* (UK), Dec 1997—Jan 1998, p. 6.

42. *Ufo* (UK), September–October 1998, p. 46.

43. *New Scientist*, 13 May 2000, p. 15.

44. Proceedings of the Second European Space Debris Conference, Darmstadt, 1997, p. 3.

45. *The Times* (UK), 16 November 1999, p. 22.

46. *The Guardian* (UK), 19 January 1999, p. 9.

47. *New Scientist*, 11 November 2000, p. 17.

48. *Scientific American*, August 1998, p. 45.

49. Proceedings of the Second Conference, p. 47.

50. *Scientific American*, August 1998, p. 46; Proceedings of the Second Conference, p. 225.

51. Proceedings of the Second Conference, p. 545.

52. *Astronomy* (UK), December 2000, p. 61.

53. Proceedings of the Second Conference, p. 509.

54. Ibid., p. 519.

55. *Frontiers* (UK), Spring 1998.

56. Proceedings of the Second Conference, p. 36.

57. Ibid., p. 18.

58. *Frontiers*, June 1999.

59. Proceedings of the Second Conference, p. 205.

60. *New Scientist*, 4 December 1999.

61. *Quest* (UK), August–September 1997, p. 47.

62. *The Times* (UK), 16 November 1998, p. 23.

63. *Unopened Files* (UK), Winter 1998, p. 9.

64. *Sunday Times* (UK), 14 November 1999, p. 10.

65. *The Times* (UK), *Interface* (suppl.), 13 May 1998, p. 11.

66. *Unopened Files*, Winter 1998, p. 45.

67. Ibid.

68. Proceedings of the Third Conference, p. 231.

69. *New Scientist*, 11 November 2000, p. 17.

70. Proceedings of the Third Conference, p. 238.

71. *The Times* (UK), 16 November 1999, p. 22.

72. *New Scientist*, 28 November 1998, p. 20.

73. Ibid., 7 November 1998, p. 25.

74. Ibid., 19 November 1999, p. 11.

75. *Tel Aviv University News*, Spring 2000.

76. *New Scientist*, 11 November 2000, p. 17.

CHAPTER 9: THE SPACE WEATHER THREAT

1. Cited in *the Times* (UK), 3 February 1997.

2. Proceedings of the Third European Conference on Space Debris, SP-473, p. 5883.

3. See *New Scientist*, 21 November 1998.

4. *Journal of Radioactivity*, 2001, vol. 53, pp. 231.

5. *New Scientist*, 14 November 1998, p. 39.

6. Ibid.

7. Ibid., p. 40.

8. *The Times* (UK), 14 October 2000, p. 5.

9. *New Scientist*, 14 October 2000, p. 5.

10. *New Scientist*, 6 January 2001, p. 11.

11. *New Scientist*, 10 April 1999, p. 23.

12. *Sunday Times* (UK), 29 November 1998, p. 1.10.

13. *International Herald Tribune*, 24 May 2001, p. 1.

14. *New Scientist*, 23 February 1996, p. 24.

15. *Sunday Telegraph* (UK), 19 July 1998, p. 4.

16. *New Scientist*, 21 May 1997, p. 31.

17. Ibid., p. 28.

18. *New Scientist*, 15 August 1998, p. 26.

19. *New Scientist*, 3 October 1998, p. 5.

20. *Sunday Times* (UK), citing George Simnet of Birmingham University, 2 March 1997.

21. *Physical Review Letters*, (UK) August 1998.

22. Proceedings of the Third Conference, p. 287.

23. *Discover*, August 1995, p. 55.

24. *New Scientist*, 27 February 1999, p. 30.

25. *The Times* (UK), 9 September 1996, p. 50.

26. *The Times* (UK), 6 August 1999, p. 16.

27. Ibid., 29 June 1999, p. 11.

28. *New Scientist*, 19 August 2000, p. 18.

29. Ibid.

30. *New Scientist*, 27 May 2000, p. 7.

31. *International Herald Tribune*, 2 February 2000, p. 2.

32. *New Scientist*, 10 June 2000, p. 7.

33. *Frontiers* (UK), Xmas 1998.

34. *Time*, 9 September 1996, p. 50.

35. *New Scientist*, 3 February 1996, p. 24.

36. *The Economist* (UK), 13 March 1999, p. 134.

37. *The Observer* (UK), 4 July 1999, p. 13.

38. *New Scientist*, 27 February 1999, p. 32.

39. "Equinox," Channel-4 productions, 1998, p. 12.

40. Ibid., p. 14.

41. *The Times* (UK), 10 November 1999, p. 9.

42. *Time*, 9 September 1996, p. 50.

43. *New Scientist*, 3 February 1996, pp. 22–26.

44. *Unopened Files*, Winter 1998.

45. *Sunday Times* (UK), 2 March 1997.

46. "Equinox," 1998, p. 98.

47. *New Scientist*, 7 November 1998, p. 5.

48. *Scientific American*, April 2001, p. 74.
49. *New Scientist*, 7 November 1998, p. 5.
50. *New Scientist*, 22 March 1997.
51. *Guardian* (UK), *Science* (suppl.), 25 November 1999.
52. *Discover* (US), August 1995, p. 61.
53. *New Scientist*, 10 June 2000, p. 7.
54. Ibid., 31 May 1997, p. 49.
55. Antony Milne, *Earth's Changing Climate*. Prism Press, 1988, p. 34.
56. *New Scientist*, 18 May 1996.
57. *Time*, 9 September 1996, p. 50.
58. *Time*, 21 February 2000, p. 62.
59. *The Times* (UK), 10 November 1999, p. 9.
60. *Frontiers* (UK), May 1999, p. 28; *New Scientist*, 22 May 1999, p. 52.
61. *New Scientist*, 1 February 1997.
62. *New Scientist*, 10 June 2000, p. 5.
63. *The Times* (UK), 10 November 1999, p. 9.
64. *New Scientist*, 19 January 2002, p. 46.

CHAPTER 10: PROTECT OUR SPACECRAFT

1. Proceedings of the Second European Conference on Space Debris, Darmstadt, 1997, p. 225.
2. Ibid., p. 25.
3. Urias Col. J.M. et al., *Planetary Defense; Catastrophic Health Insurance for Planet Earth*. Springer-Verlag, 1998.
4. Proceedings, p. 132.
5. Ibid., p. 129.
6. Ibid., p. 9.
7. *Journal of Radioactivity*, 2000, vol. 53, pp. 231.
8. Proceedings of the Second Conference, p. 183.
9. *Scientific American*, August 1998, p. 45.
10. *The Guardian* (UK), 19 January 1999, p. 9.
11. Proceedings of the Second Conference, p. 285.
12. Ibid., p. 201.
13. Proceedings of the Third European Conference on Space Debris, SR473, p. 15.
14. Orbital Debris Quarterly News Online, vol. 7, issue 1.
15. Proceedings of the Second Conference, p. 139.
16. Proceedings of the Third Conference, p. 203.
17. *The Economist* (UK), 25 March 2000, p. 118.
18. Proceedings of the Second Conference, p. 202.
19. *Quest* (UK), April 1999, p. 10.
20. Proceedings of the Second Conference, p. 285.
21. *The Guardian* (UK), 19 January 1999, p. 9.

22. *Astronomy* (UK), December 2000, p. 6.
23. *New Scientist*, 4 August 2001, p. 21.
24. Proceedings of the Third Conference, p. 304.
25. Ibid., p. 705.
26. Proceedings of the Second Conference, p. 28.
27. *Scientific American*, p. 47.
28. Proceedings of the Second Conference, p. 38.
29. *New Scientist*, 29 November 1997, p. 7.
30. *The Economist* (UK), 28 August 1999, p. 70.
31. *New Scientist*, 30 October 1999, p. 22.
32. *Alien Encounter* (UK), December 1997, p. 37.
33. *New Scientist*, 1 April 2000, p. 61.
34. *Scientific American*, August 1998, p. 47.
35. *New Scientist*, 31 July 1999, p. 7.
36. Jack Challoner, *Space*. Channel-4 Books, 2000, p. 109.
37. *New Scientist*, 19 August 2000, p. 11.
38. *The Times* (UK), 28 October 1996.
39. *New Scientist*, 19 August 2000, p. 11.
40. *New Scientist*, 24 June 1995.
41. *The Times* (UK), 28 October 1996.
42. *New Scientist*, 19 August 2000, p. 11.
43. *Sunday Times* (UK), 18 June 2000, p. 16.
44. Proceedings of the Second Conference, pp. 18–19.
45. *New Scientist*, 29 November 1997, p. 7.
46. *New Scientist*, 20 January 2001, p. 21.
47. *New Scientist*, 14 November 1998, p. 41.
48. *The Guardian* (UK), 19 January 1999, p. 9.
49. *Focus* (UK), November 2000, p. 110.
50. *High Performance Polymers*, vol. 13, September 2001, p. 401.
51. *Sunday Times Culture* (UK), 21 November 1999, p. 62.
52. *Advanced Materials*, vol 2, p. 957; *New Scientist*, 28 August 1999, p. 12.
53. *New Scientist*, 11 April 2000, p. 11.
54. *New Scientist*, 14 November 1998, p. 40.
55. *New Scientist*, 11 April 2000, p. 11.
56. *Scientific American*, August 1998, p. 47.
57. *Sunday Times* (UK), 9 May 1999, p. 5.11.
58. Ibid.

CHAPTER 11: DEFEND THE EARTH

1. Austen Atkinson, *Impact Earth*. Virgin Publishing, 2000, p. 136.
2. *The Times* (UK), letter, 25 September 2000, p. 17.
3. Ibid.
4. *Time*, special issue, Winter 1997–1998, p. 64.

5. *The Times* (UK), 14 November 1993.

6. Gerrit Verschuur, *Impact: The Threat of Comets and Asteroids.* Oxford University Press, U.S., 1996, p. 206.

7. *Time* 15 February 1993.

8. *New Scientist*, 21 March 1998, p. 5.

9. *Time* 15 February 1993.

10. *Sunday Times* (UK), 14 March 1996.

11. Verschuur, *Impact*, p. 209.

12. *New Scientist*, 14 November 1992, p. 4.

13. *Time*, 23 March 1998, p. 89.

14. Verschuur, *Impact*, p. 210.

15. *Time*, special issue, Winter 1997–1998, p. 69.

16. Verschuur, *Impact*, p. 205.

17. *Time*, 23 March 1998, p. 89.

18. *The Times* (UK), 26 May 2001, p. 8.

19. Jerry Smith *HAARP: The Ultimate Weapon of the Conspiracy.* Adventures Unlimited Press, 1998, p. 18.

20. *Ufo Reality*, citing Dr. Nick Begich, November 1997, p. 44.

21. Smith, *HAARP*, p. 16.

22. Ibid., p. 18.

23. *The Times* (UK), 14 March 1993.

24. *New Scientist*, 13 November 1995.

25. *The Times* (UK), 13 March 1998, p. 5.

26. Atkinson, *Impact Earth*, pp. 155–56.

27. *Time*, 15 February 1993.

28. *New Scientist*, 13 November 1995.

29. Atkinson, *Impact Earth*, p. 145.

30. *International Herald Tribune* 18 October 2000, p. 1.

31. *Spaceguard (UK), Impact*, bulletin no. 11.

32. Ibid.

Further Reading

BOOKS

Atkinson, Austen. *Impact Earth*. Virgin Publishing, 1999.

Challoner, Jack. *Space*, Channel-4 Books, 2000.

Davies, James. *Space Exploration*. Chambers, 1992.

Irvine, Matt. *Tele-Satellites*. Gloucester Press, 1989.

Urias, Col. J. M. et al., *Planetary Defense: Catastrophic Health Insurance for Planet Earth*, 1998.

Proceedings of the Second Conference on Space Debris, Darmstadt, SP-393, 1997.

Verschuur, Gerrit. *Impact: The Threat of Comets and Asteroids*, Oxford University Press, 1996.

JOURNALS AND MAGAZINES

Advanced Materials, vol. 2, p. 957.

Astronomy (UK), December 2000, p. 56.

Frontiers (UK), November 1998, p. 39.

Frontiers (UK), December 1998, p. 75.

High Performance Polymers (UK), September 2001, vol. 13.

Journal of Radioactivity, 2001, vol. 53, p. 231.

"The Leonid 1998 Shower: Information for Spacecraft Operators," European Space Agency, Mission Analysis section, 1998.

New Scientist (UK), 7 November 1998, p. 5.
Physical Review Letters (UK), August 1998.
Quest Publications (UK), series of QP publications on the Moondust Project, 1998.
Scientific American, August 1998, p. 44.
Spaceflight, March 1995, vol. 37.
Impact Spaceguard (UK), bulletins 9 and 11.
Tel Aviv University News, Spring 2000.
Ufo (UK), March–April 1999, p. 34.
Ufo (UK), July–August 1997.
Unopened Files (UK), December 1999, p. 59.

Index

Index

About the Author

ANTONY MILNE is an associate fellow with Spaceguard UK and was formerly an analyst with NATO's Scientific Affairs Division. Among his five earlier books is *Doomsday: The Science of Catastrophic Events* (Praeger, 2000).